Introduction to

FIELD ARTILLERY ORDNANCE 1861-1865

To Homer McCall,
Hope you enjoy my book.
Best wishes,

by Jack W. Melton, Jr., and Lawrence E. Pawl

Kennesaw Mountain PRESS

Kennesaw Mountain Press, Inc., Kennesaw, Georgia
1810 Old Highway 41, Kennesaw, GA 30144

Printed in the United States of America. First edition.

The authors invite comments, corrections, and additional information from the readers. New information is most appreciated and will be acknowledged in future editions of this book. Correspondence can be sent to Kennesaw Mountain Press, Inc., at the address above.

Additional copies of this title can be ordered for $24.95 plus $3.50 shipping and handling from Kennesaw Mountain Press, Inc., at the address above.

Library Of Congress Cataloging in Publication Data

Melton, Jack W., Jr., and Pawl, Lawrence E.
Introduction to field artillery ordnance 1861-1865.
93-077319

ISBN 0-9635861-4-9

Typography and design by
Nancy Dearing Rossbacher
Fit To Print Desktop Publishing Services
Post Office Drawer 729, Orange, Virginia 22960

This volume is dedicated to the memory of

Thomas S. Dickey, Sr.

who was a prominent Civil War historian and the premier Civil War artillery projectile collector of this century. Mr. Dickey's accomplishments included serving as president of the Atlanta Civil War Roundtable and co-authoring three books on Civil War artillery projectiles. His personal artillery collection, currently under the control of the Atlanta History Center, is legendary in its size and scope and features many one-of-a-kind items.

Tom was a true southern gentleman and a giant in the Civil War relic collecting field. He is greatly missed by all his fellow collectors.

—*The Authors*

Acknowledgments

The authors sincerely appreciate the following individuals, who graciously offered their expertise and knowledge, and without whom this work would not have been possible. Special thanks are extended to Chuck Jones and Steve Phillips for the use of their collections. Editorial assistance from Henry Higgins was especially appreciated.

Howard Alligood
Lon Ashbrook
Ray Basco
Jack Bell
Chris Bullington
Brant Calvit
Michael Carpenter
Allen Casey
Rex Cockerham
Dennis Cox
Ray Davenport,
Sumter Military Antiques
David Davidson
Dot Dickey
Jerry Fertitta,
The Richmond Arsenal
Dr. Jerry Foster
Stanley Franklin
Tim Garrett, The Picket Post
N. Gibson Jr.
Robert Glessner
Greg Griffin
Bob and Linda Hammonds
Charles Harris
Bill Henderson, The Picket Post
Larry Hicklen
The Horse Soldier of Gettysburg
Barry Hocutt
Corkey Huey
Jerry Imperio
Tammy Johnson
C. Peter Jørgensen, *The Artilleryman*
Mike Klinepeter

Every successful work is, in the final analysis, a collaborative effort.

Glen E. Mattox
Steve Mullinax
R.E. Neville, Jr.,
Military Antiques
Michael J. O'Donnell
Denny Pizzini
Daniel Poppen
Lee Poss
Steve Ransbotham
Harry Ridgeway
Andy Rice
Warren Ripley
Bruce Roberts
Peter Rockefeller
Nancy Dearing Rossbacher,
North South Trader's CIVIL WAR
and Fit To Print Desktop Services
Lee Sanzo, The Galvanized Yankee
Carl Sitherwood
Frank Smith
Smithsonian Institution
Stone Mountain Relics, Inc.:
John Sexton, Charles and Nan Nash,
and Lori Nash Cosgrove
Stephen W. Sylvia, Publisher's Press, Inc.
Dean Thomas, Thomas Publications
Joe Troxtel
Warren Vitellaro
Ronnie Webb
Don Williams
Ken Williams
George F. Witham
F. Paul Wohlford

The authors wish to extend special thanks to Nancy Dearing Rossbacher of Fit To Print for her graphic design throughout this work.

Co-author Jack Melton, Jr., would like to express his gratitude to Lawrence Pawl, whose enthusiasm and encouragement were the driving forces behind this work.

Preface

During the American Civil War, more varieties of projectiles and cannon were used than in any other time in military history. The outbreak of hostilities in 1861 found inventors on both sides searching for the perfect blend of sabot, shell body, and fuse to create the artillery projectile that would give the military advantage to their respective cannoneers. This seemingly unending search for that elusive, perfect projectile continued until the end of the war.

As a result of these countless innovations, the student of Civil War ordnance today faces a fascinating and potentially confusing maze of battlefield-tested artillery projectiles.

This volume has been written for the purpose of helping the reader to better understand the variety of ordnance used by the Confederate and Federal forces. This book is not intended to be—nor should it be used as—the absolute final word on this fascinating area of research. The authors have made every effort to be accurate, and whenever possible, archival and patent records have been used to support the information contained herein.

Civil War ordnance offers a fascinating and seemingly limitless course of study.

The terminology associated with the study of Civil War artillery can be technical, and we have prepared an extensive glossary to assist the reader. This can be found on pages 10-13, and the reader may want to refer to it when encountering unfamiliar terms in this book.

This work also features several innovations in the presentation of Civil War projectiles. The reader should note that we have prepared a section on non-combat-proven / post-war projectiles (pages 237-257). This is designed to make the student of Civil War artillery aware of projectiles that could be confused with legitimate, proven Civil War examples.

Another innovation is a collector's guide to determining the relative value of a projectile (pages 258-259). Others can be found in the appendixes, which include an explanation of the often confusing caliber to pounder relationship (pages 260-261) and recommendations concerning the disarmament and preservation of artillery projectiles (pages 270-272).

Those readers who desire further detailed information in this vast field of study will find a recommended reading list in Appendix E (page 273).

The study of Civil War projectiles is an ongoing endeavor, and we welcome comments and additional information from our readers.

—The Authors

Table Of Contents

I. Basic Facts Concerning Field Artillery

The use of field artillery has always had a demoralizing effect on soldiers. Throughout the Civil War, the tactical use of field artillery evolved as the quality of field ordnance improved. Nothing was more frightening to the infantry of both sides than having to charge against emplacements of cannon firing shot, shell, and canister at them with devastating precision.

THE CONFEDERACY & BRITISH IMPORTS

At the outbreak of hostilities in 1861, most of the ordnance and cannon the Confederates States had at its disposal were in the captured Federal armories and fortifications. Unfortunately, the quantities of ordnance were not great. Most accounts place the initial Confederate cannon cache at 35 field artillery pieces. This information is found to be inaccurate: it refers only to field artillery pieces and only the ones captured in forts taken over by the Confederates. Alabama alone seized approximately 100 cannon (24- and 32-pounders) in its seacoast forts on January 15, 1861, even before the state seceded.

Standardization of cannon caliber did not occur until near the end of the Civil War.

Since initially the South only had one functional cannon foundry (Tredegar Iron Works in Richmond, Virginia) with which to supply its armies, outside sources were needed to meet demands. Great Britain was one of the many sources of projectiles and cannon used by the Confederacy. Of these British imports, the Whitworth system is the most recognized and desirable today due to its unusual bore-shaped projectiles. Other British imports included the Armstrong, Britten, and Blakely systems of cannon and projectiles.

RIFLED ORDNANCE vs. SMOOTHBORE

The ordnance available early in the war was approximately one-third rifled, with smoothbore (6- and 12-pounder cannon) making up most of the remaining two-thirds. As of December 1, 1861, there were only 100 wrought-iron (ordnance) rifles, 193 Parrott cannon of both field sizes on hand, and only 36 Model 1857 Napoleon cannon for the Federal artillery. The rest of the Federal Army was equipped with Model 1841 6-pounders, heavy 12-pounders, and 12-pounder field howitzers—all smoothbores. The younger artillery officers were impressed with the rifled system, while the older, more experienced officers maintained that the smoothbore cannon was equally as important.

As the war progressed, many battles occurred at close range, under which condition the 12-pounder canister proved to be both deadly and defensive. The rifled artillery proved to be better for offense because of their accuracy in counter-battery work and also the affect of long-range shell and case-shot.

The continued success of this system was evidenced by the fact that by the cessation of hostilities in 1865, fully half of the field artillery pieces used by both North and South were 12-pounders.

EXPERIMENTATION LED TO DIVERSITY

Innovations in artillery development were slow to gain acceptance in the United States during the first half of the 19th century. Inventors spent years on experimentation, field trials, obtaining financial support, and engaging in the inevitable political maneuvering. All of this was necessary to have any newly developed artillery projectile or cannon reviewed by the Federal Ordnance

Department for possible military approval.

The outbreak of hostilities in 1861 changed much of this. Developments in the field of artillery reached a near fever pitch in both the Federal and Confederate war departments. Robert Parrott of Cold Spring, New York, and John Read of Tuscaloosa, Alabama, were two pre-war pioneers in the design and production of rifled artillery projectiles and cannon. Read offered his services to the Confederacy, while Parrott remained loyal to the Federal government. These two inventors produced more artillery patterns and varieties of projectiles than any other inventors during that period.

All this activity led to rapid diversifications in field artillery projectiles and cannon systems. One can imagine the problems encountered by an army that had to carry 2.9-inch, 3-inch, 3.67-inch, and 4.62-inch caliber projectiles for each of the different cannon systems they had in the field. Standardization of cannon caliber was desperately needed, but it was not until near the end of the war that the 3-inch rifle, 12-pounder, and 20-pounder cannon became the unofficial standards for field artillery use.

The authors know of more than 1,000 different artillery projectiles tested on the battlefield from 1861 to 1865. It is this great variety of projectiles that continues to challenge and excite the collector and student of Civil War artillery today.

II. Field Artillery Projectiles

A. Glossary

BLIND SHELL: A projectile with a hollow cavity that was plugged and therefore not intended to have been fitted with a fuse. The projectile weighed less than a solid shot of the same caliber and thus had increased velocity when fired.

BOLT: An elongated projectile, usually made of solid iron, without a cavity or fuse. Most often a rifled projectile.

BORE DIAMETER: The cannon diameter at the muzzle measured from wall to wall in a smoothbore and from land to land in a rifled tube.

To understand the projectiles, it is necessary to understand the terminology.

BOURRELET: A French term referring to raised rings that are larger in diameter than the projectile body. These were used as bearing surfaces for the projectile. They were turned on a lathe to the dimension needed so that the projectile would fit into the bore of the cannon.

BURSTING CHARGE: The gunpowder charge that is contained inside the cavity of the projectile.

CALIBER: The diameter of a cannon's bore expressed in inches, or the weight of the solid shot corresponding to it. The inside diameter of a firing tube (cannon, howitzer, mortar, etc.).

CANISTER: A cylinder made of tin, iron, or sometimes lead, with a removable thin iron top. A heavy iron plate was most often located between the canister balls and the wooden sabot at the bottom. The cylinder had metal balls, most often made of lead or iron, which were arranged in rows. Commonly sawdust was used to fill the space between these balls. The edges of the vertical cylinder walls were bent over the iron top to help keep the canister contents in place. Canister was designed to be used in close range against enemy troops.

CARTRIDGE BAG: The black powder propellent charge for the projectile was placed in this bag. Commonly made of a merino, wildbore, or bombazette wool material. Sometimes referred to as a powder bag.

CASE SHOT: Similar to the common shell except that the walls of the projectile are thinner. In both spherical and rifled projectiles, the bursting charge was usually located in a thin metal container, commonly made out of tin or iron. The case-shot material was filled around this container. The internal cavity was usually filled with lead or iron balls in a sulphur or pitch matrix. The small bursting charge of black powder was designed to disperse the case-shot balls in a cone-shaped pattern. The concept was to give the same effects of canister, but at much longer ranges. Case shot was invented by an Englishman, Lieutenant Henry Shrapnel of the Royal Artillery, in 1784. See page 26, figure II B-12, for an example of a cross-section of a case-shot projectile.

COMBINATION FUSE: A fuse that was designed to automatically ignite, without needing a flame. If the initial discharge of the cannon did not ignite the fuse, then the fuse was designed to explode on impact.

CONCUSSION FUSE: A fuse that was designed to explode with the projectile in any impact position or angle of impact. Mostly found in spherical projectiles.

EARS: Indentations found on mortar projectiles. Designed to aid in loading and aligning the projectile so the fuse was in the center of the mortar's bore.

FIXED AMMUNITION: Most often a pre-assembled combination of smoothbore projectile, sabot, and powder bag. This assembly allowed for an increase in the rate of fire of the gun crew.

FLAME GROOVES: Grooves cut or cast into the body and/ or the sabot of the projectile. These improved the chances of the flame from the firing charge successfully passing over the projectile and lighting the time fuse.

FLANGE: A projecting rim or ridge on a projectile's body for the purpose of guiding it along the grooves of the cannon, thus imparting rotation.

FUSE: A device screwed or hammer-driven into the body of a projectile. It was designed to ignite the gunpowder bursting charge.

FUSE PLUG: Sometimes called a fuse adapter. A metal or wood device that was screwed or driven into the projectile body to hold a paper time fuse.

GATE: The waste piece of metal cast in the opening or passage through which molten metal is poured or run into a mould.

HOT SHOT: A solid iron projectile that was intended to be heated red hot. In some cases, it was separated from the powder charge by an iron base that acted as a sabot. Its purpose was to create fires in enemy fortifications.

LATHE DIMPLE: A drilled or countersunk depression found on the base or sabot of rifled projectiles. This is the true center of the projectile and the site where the lathe arbor held the projectile in place while it was being turned on a lathe.

LATHE DOG: Found on the ogive or nose of rifled projectiles; often referred to as a lathe lug or notch; a raised area on the body of the projectile that kept it from slipping while being turned to bore tolerance on a lathe.

MATRIX: A mixture of sulphur or pitch added to the interior of a case-shot projectile. It was used as a filler to stabilize case-shot balls in order to prevent accidental damage to the projectile's bursting charge can or cavity. It also prevented a premature discharge of the bursting charge by movement of the case-shot balls during excessive projectile motion.

MOULD VENT: A small opening in the mould for the release of gases, thus allowing the molten metal to completely fill the mould.

NIPPLE: A fitted device attached to the surface contact end of a percussion fuse slider; designed to support a percussion cap. Similar to a nipple or cone on a Civil War rifle.

OGIVE: The curve that determines the shape of the nose of a conical projectile.

PAPER TIME FUSE: A tapered, gunpowder-based resin cone wrapped in paper. It was manufactured into standard lengths according to the desired seconds of burn needed. Ignition of the fuse took place at the time of the cannon discharge when the flames traveled past the projectile body to ignite the fuse in the fuse plug.

PERCUSSION CAP: A cup-like object usually made of copper or brass containing fulminate of mercury. It sat on the nipple end of a fuse slider used with the percussion fuse. On impact, the percussion cap would send a spark to the bursting charge.

PERCUSSION FUSE: A fuse designed to communicate fire to the bursting charge of the projectile at the moment of impact with the ground or other hard surface.

PLUNGER: Sometimes called a striker or slider, this was the inside moveable part of a percussion fuse. Its purpose was to make contact with the percussion cap on the nipple and the anvil cap when the projectile struck a hard surface. This helped direct the discharge flame to the interior black powder charge.

POLYGONAL CAVITY: A method of casting the interior of a common spherical shell to improve the fragmentation. The interior was cast with lines of weakness in various patterns.

POWDER TRAIN: A channel in the interior of the projectile body, usually a metal tube; intended to increase the chances of the detonation charge reaching the bursting charge.

RIFLE: Refers to a cannon tube that was cut with lands and grooves. This was done to improve the accuracy of the rifled projectile.

RIFLING: Often referred to as lands and grooves; a firing tube (cannon) with spiral grooves cut into the bore. The expanding sabot of a fired projectile would engage the grooves for greater accuracy and stability. After groove cutting, those portions of the original smooth bore that remained are called lands and are the actual diameter of the bore.

SABOT: A device made of wood, brass, copper, lead, papier-mâché, leather, rope, or wrought iron. It served as the driving band for the projectile and was forced into the grooves of the rifling of the cannon. This caused rotation, which extended the range and improved the stability of the projectile. A wooden sabot was used in a smoothbore cannon to hold the projectile with the fuse forward and in the center of the bore. Often the wooden sabot had one or more grooves on the base to which the cartridge bag was tied.

SEGMENTED SHELL: A concept similar to the polygonal-cavity spherical shell, but used in rifled projectiles. The interior of the shell is cast with lines of weakness.

SHELL: A hollow projectile filled only with a bursting charge of blackpowder. Sometimes called a common shell.

SIDE-LOADER: A hole in the side of the body of a projectile other than the fuse opening. This facilitated the insertion of the case-shot material and matrix. It was usually sealed after loading with a lead, iron, brass, or copper threaded plug.

SOLID SHOT: A solid iron projectile without the ability to contain a bursting charge or a fuse. Commonly called a bolt in rifled ordnance. Designed to knock down fortifications, earthworks, or to be fired at lines or columns of troops.

SPRUE: The overflow of metal poured through the gate in a mould to make the casting.

STAND OF GRAPE: Often referred to as grapeshot; consisted of six to nine small cast-iron balls of a size appropriate to the caliber of the cannon bore, arranged in two or three layers of three balls in each layer. Most of the Civil War stands were held together by two iron rings and a thick iron plate at each end. A rod in the center was bolted to hold the two plates together. A rope handle fitted into two holes in the top iron plate. Some stands, called quilted grape, were held together by a covering of canvas and a network of twine or wire. Stands of grape were mostly used by naval or seacoast artillery and were rarely used in the field.

TIME FUSE: A fuse designed to explode the projectile a designated number of seconds after the projectile was fired from the cannon or mortar. It was usually fired with a case-shot projectile intended for air burst over troops.

UNDERPLUG: An iron, brass, or copper circular threaded plug with a hole in its center, smaller than the Bormann fuse. Found in spherical projectiles under the Bormann fuse. This was used to help support the soft metal Bormann fuse from being deformed during firing. See page 16, figure II B-2, for an example.

VENT: The channel in a cannon through which the friction primer ignites the propellent charge for the projectile.

WINDAGE: Refers to the difference between the bore diameter and the diameter of the projectile.

II. Field Artillery Projectiles
B. Cross-Sections

The projectiles in this section have been cut in half by professionals. Under no circumstances should this be attempted by anyone without the proper training and equipment necessary to perform this life-threatening task safely. The authors do not encourage the reader to perform this work and do not hold themselves responsible for anyone attempting to do so.

This section is intended to give the reader a better understanding of the construction, design, and purpose of Civil War field artillery projectiles. Various elements of the projectiles will be numbered and descriptions provided. In addition, the reader will be referred to other sections for more information. A review of the glossary section may be helpful in understanding the terms used.

Please: Do not attempt to cross-section a projectile without the proper equipment and training.

Fig. II B-1

Federal

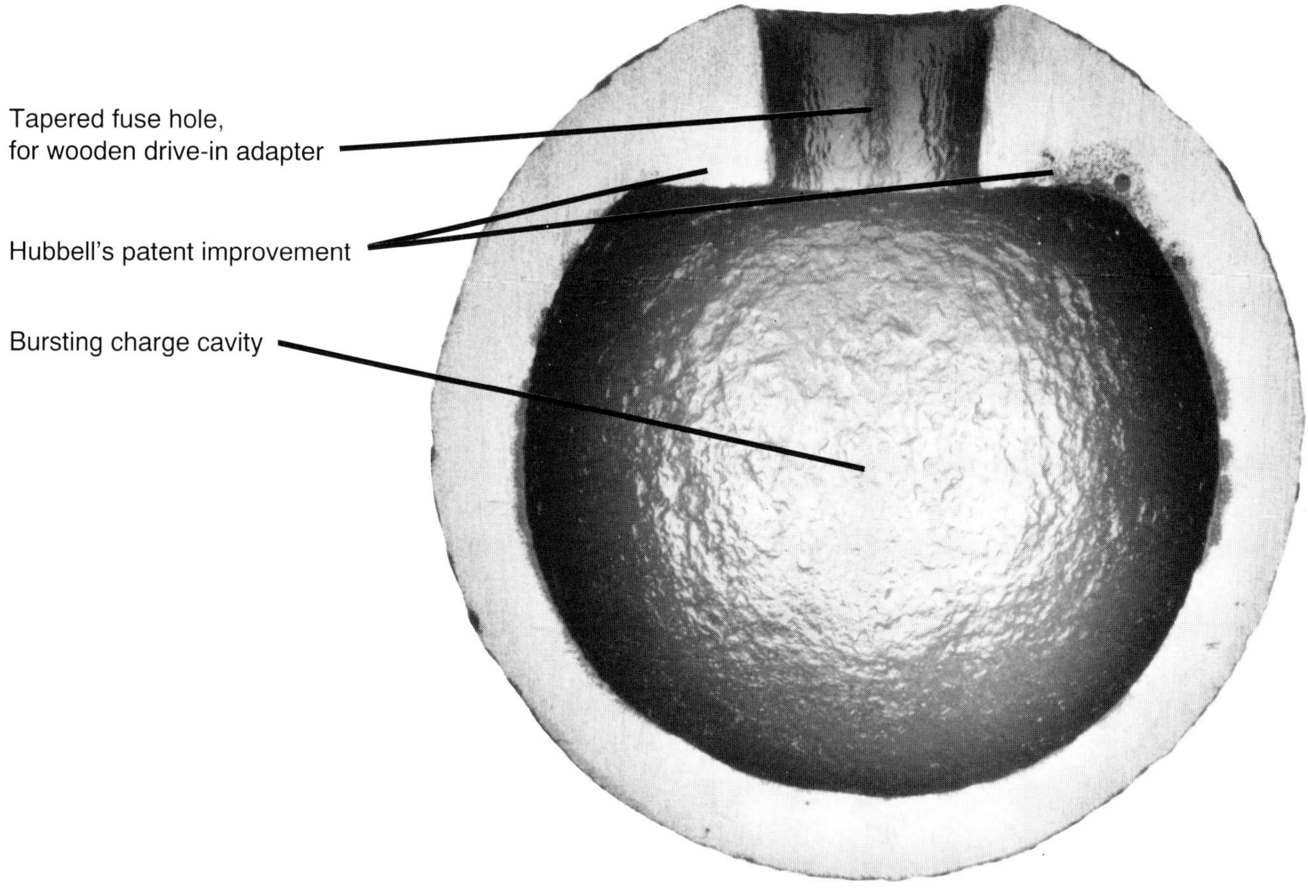

Comments: William W. Hubbell of Philadelphia, Pennsylvania, patented this projectile pattern on January 22, 1856, patent #14,133. The projectile wall around the fuse opening was thicker and flat on the bursting cavity side. This allowed for greater stability in flight and therefore more accuracy. Hubbell patented the pictured projectile as an improvement on the March 2, 1858, patent #19,505.

Fig. II B-2

Confederate

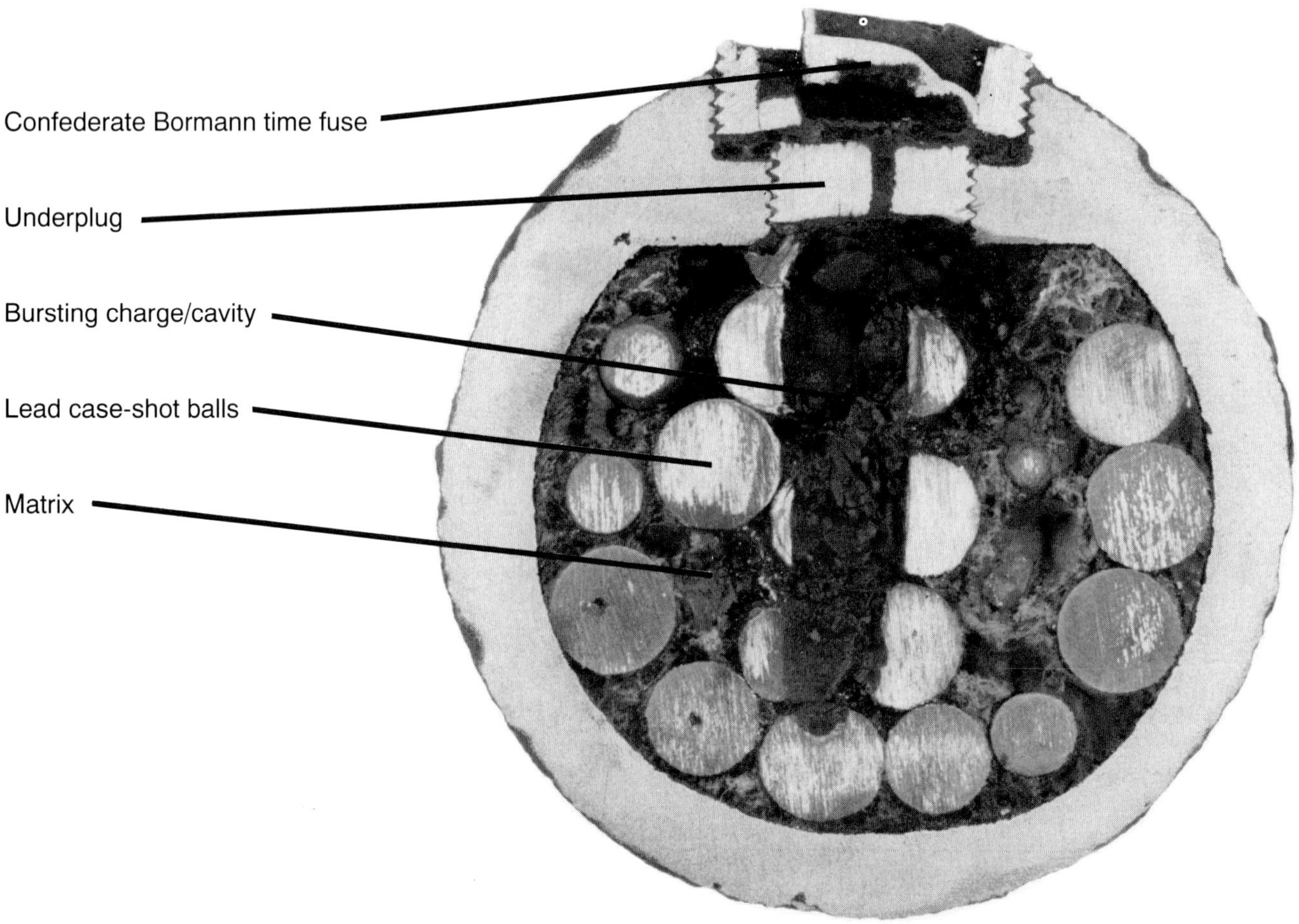

Comments: In most cases the Federal spherical case-shot projectiles contained the bursting charge in a metal cylinder as opposed to those of the Confederates, who almost always drilled into the case shot to form the bursting charge cavity. Note the lead balls, which were cut by the drill. Case shot extended the range of canister as an anti-personnel device.

Fig. II B-3

Confederate

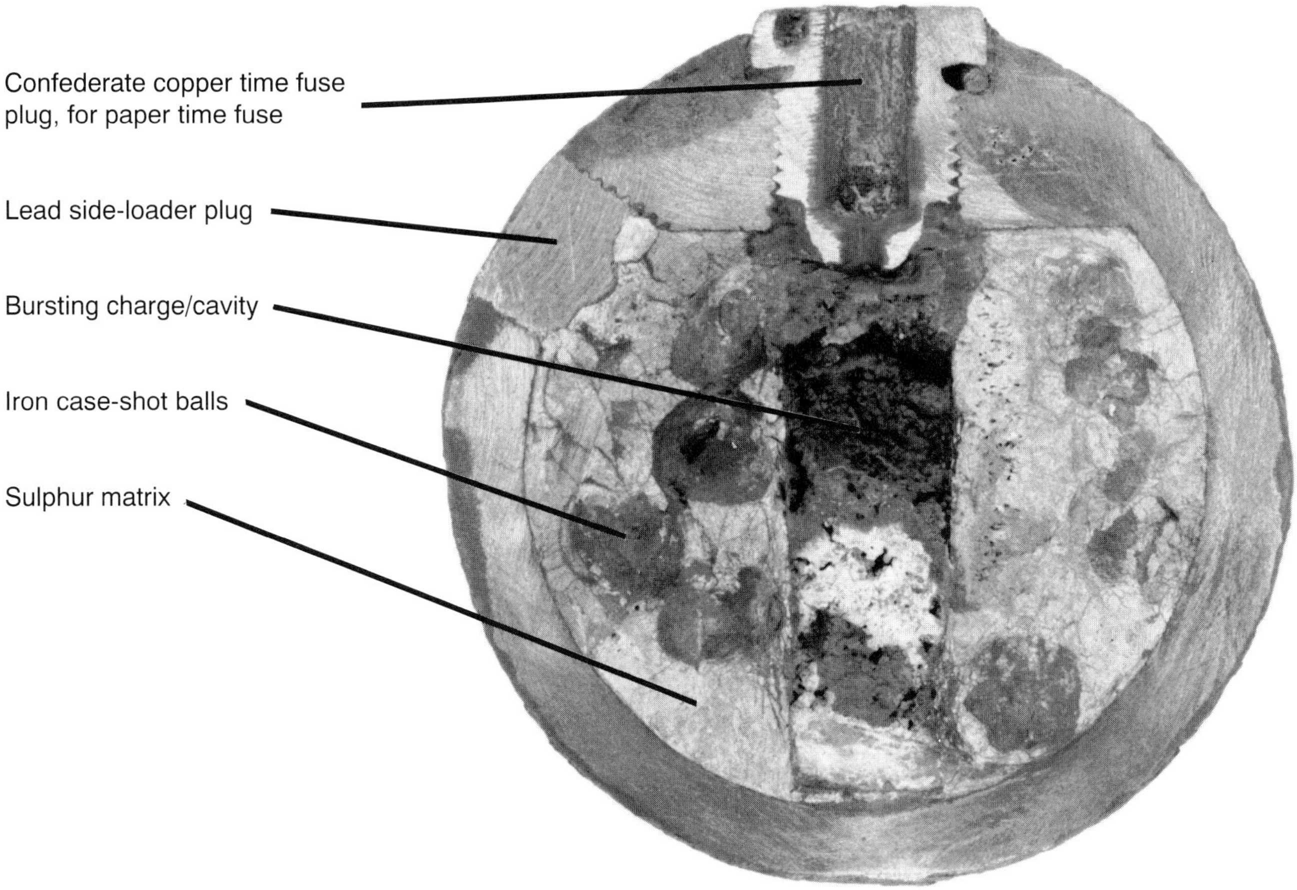

Comments: Due to the shortage of lead needed by the Confederates for small arms ammunition, iron case-shot balls were substituted for lead. This created a problem in that when the Confederates went to drill into the iron case shot, the drill bit could not penetrate the hardened balls easily and would often bind, causing the matrix to break loose. This problem was remedied by the use of a dowel. The process used to construct the pictured specimen is as follows: After the casting was removed from the mould, the fuse and side-loader holes were finished and threaded, a dowel was inserted into the fuse hole to the opposite side, and the iron balls were inserted through the side-loader hole, followed by hot matrix. After the matrix cooled the dowel was removed, leaving a space for the bursting charge.

Fig. II B-4

Federal

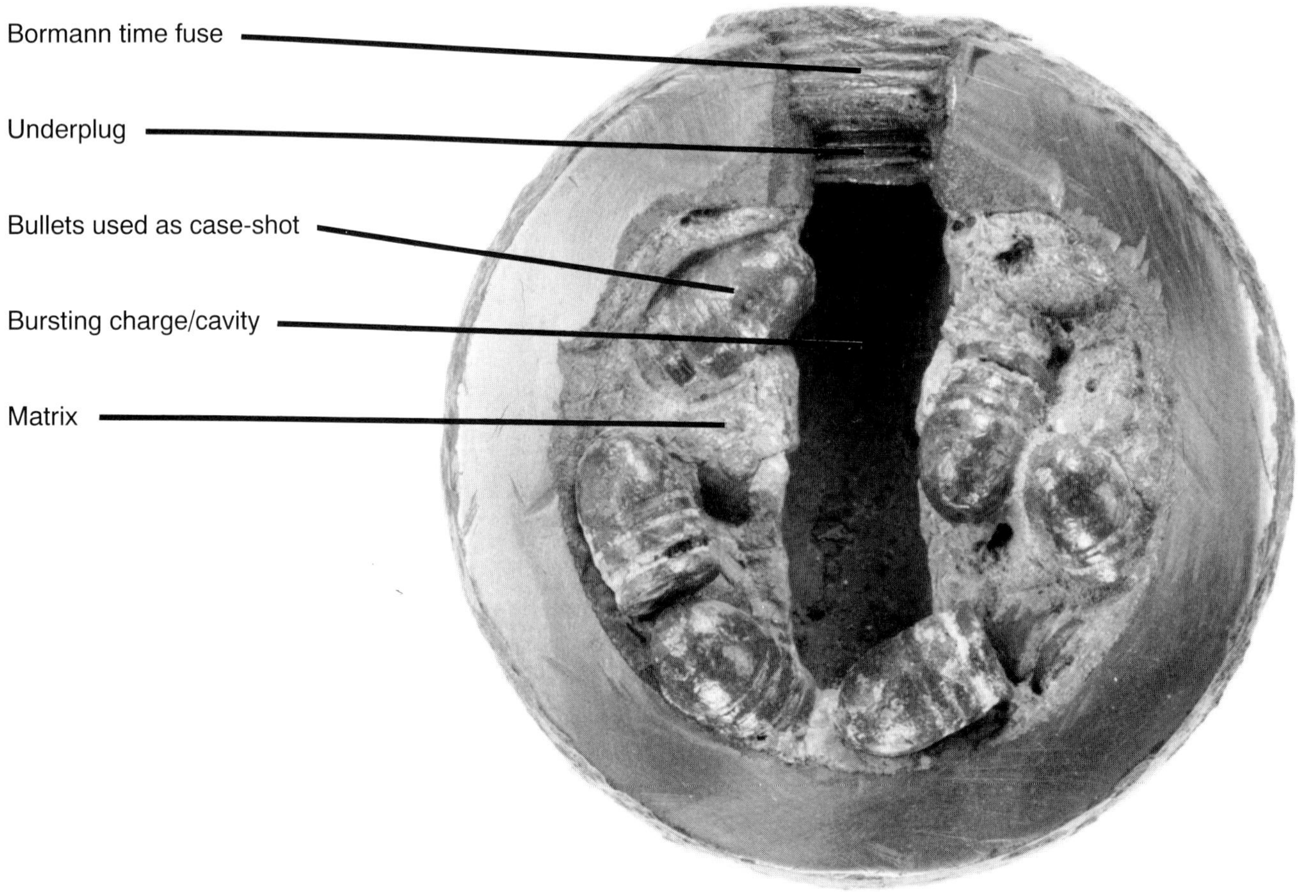

Comments: The Federals and Confederates occasionally used bullets for case-shot filler. This is an example of the Federals having used .58 caliber Williams cleaner bullets for this purpose.

Fig. II B-5

Confederate

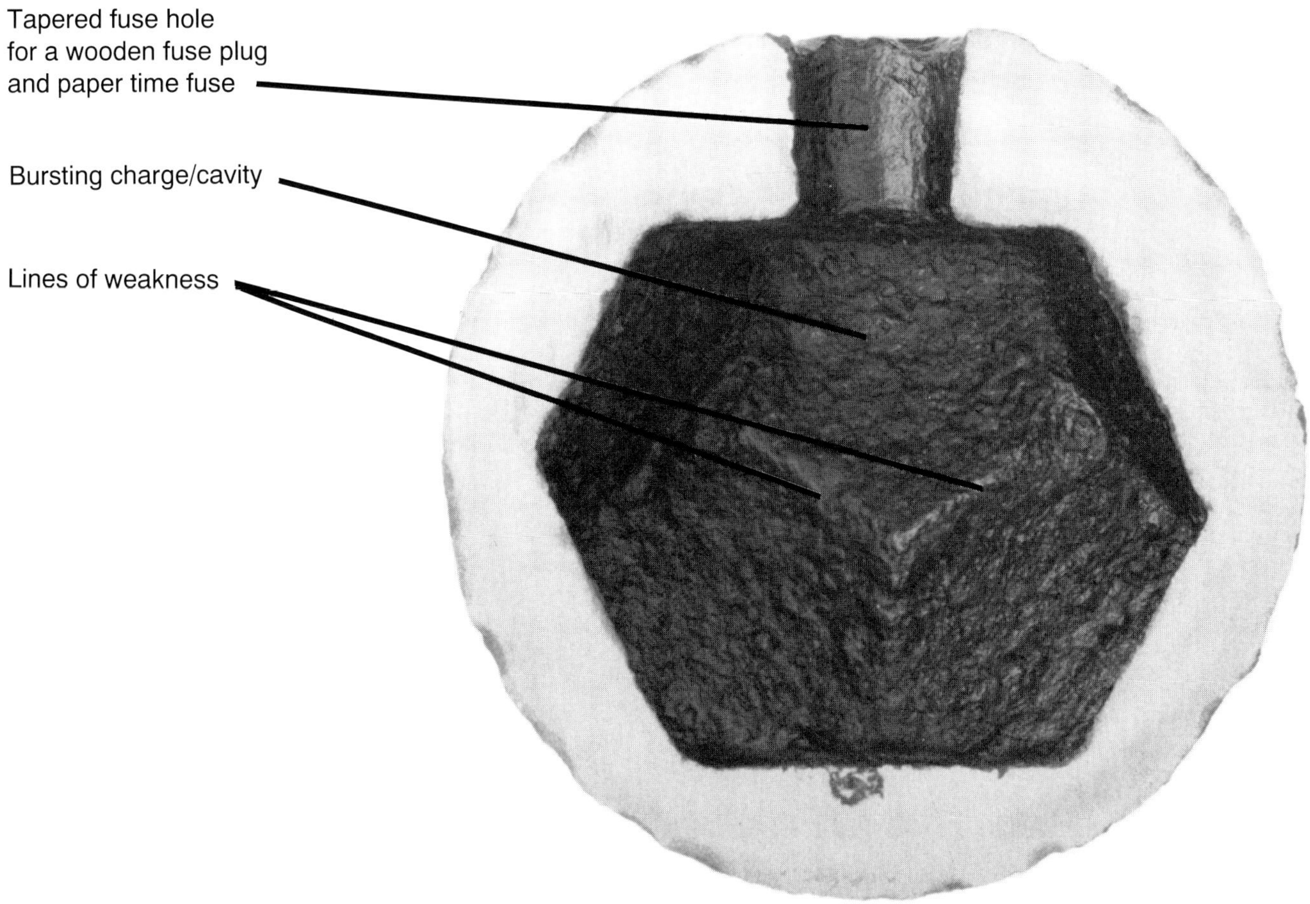

Comments: The concept of polygonal-cavity projectiles is generally attributed to Confederate Colonel John W. Mallet of Macon, Georgia ,whose ideas may have been influenced by the segmented projectiles of British origin. There are at least four different designs of polygonal cavities known to the authors. The inner casting core created lines of weakness, which aided in the creation of a uniform number of fragments when ignition of the bursting charge fractured the projectile. This style of projectile was popular with the Confederate forces, which used it heavily in place of case shot. See page 57, figure III A-10, for a complete projectile.

Fig. II B-6

Federal

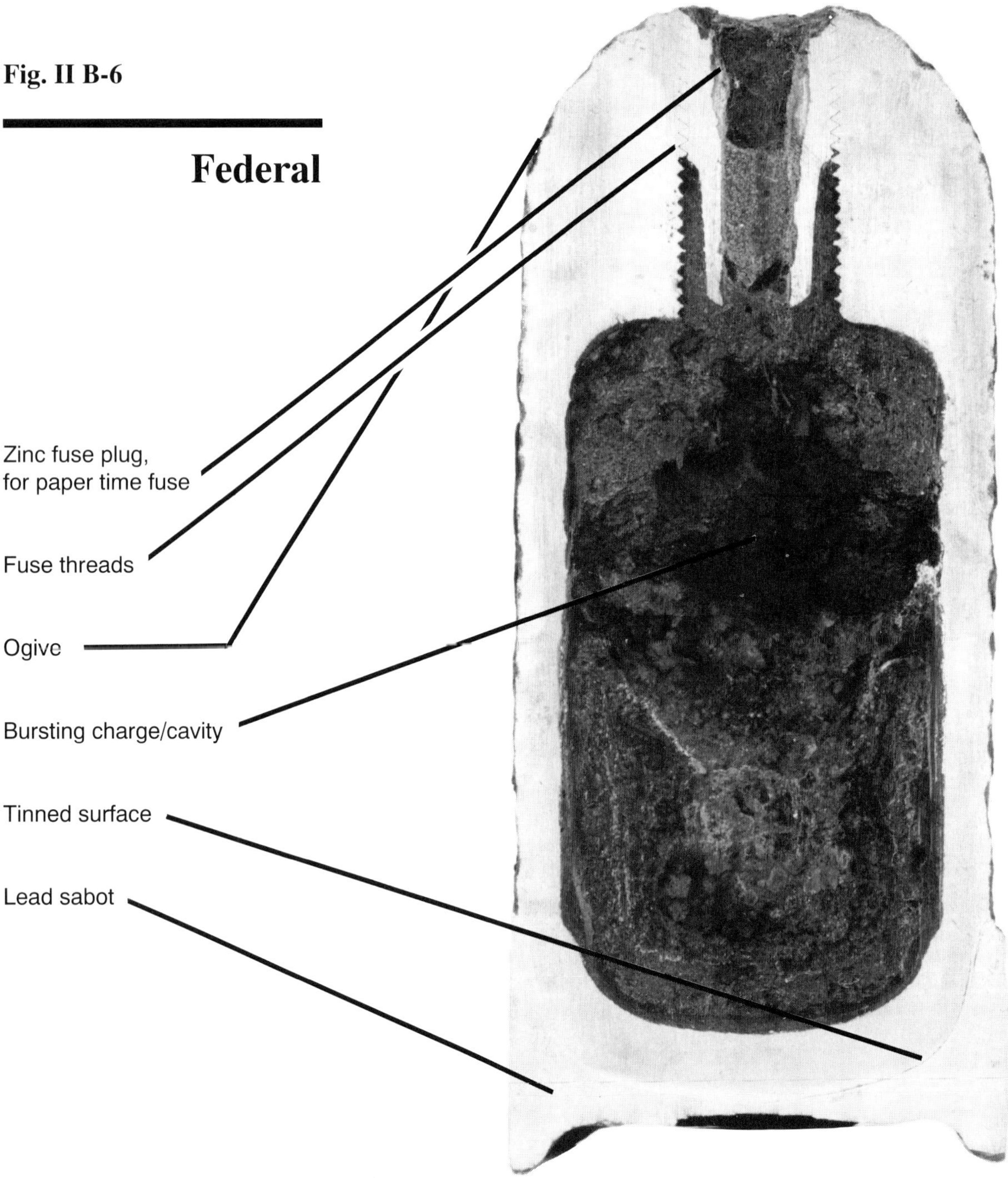

Comments: This pictured specimen is a Federal Dyer that has been misidentified in the past as a Confederate Burton projectile. However, located in the National Archives in Washington, D.C., and in *The War of the Rebellion* (Official Records) is information that supports the conclusion that the projectile is of Federal manufacture. Note the tinned surface between the iron base of the projectile and the lead cup. See page 112, figure III A-65, for further comments.

Fig. II B-7

Federal

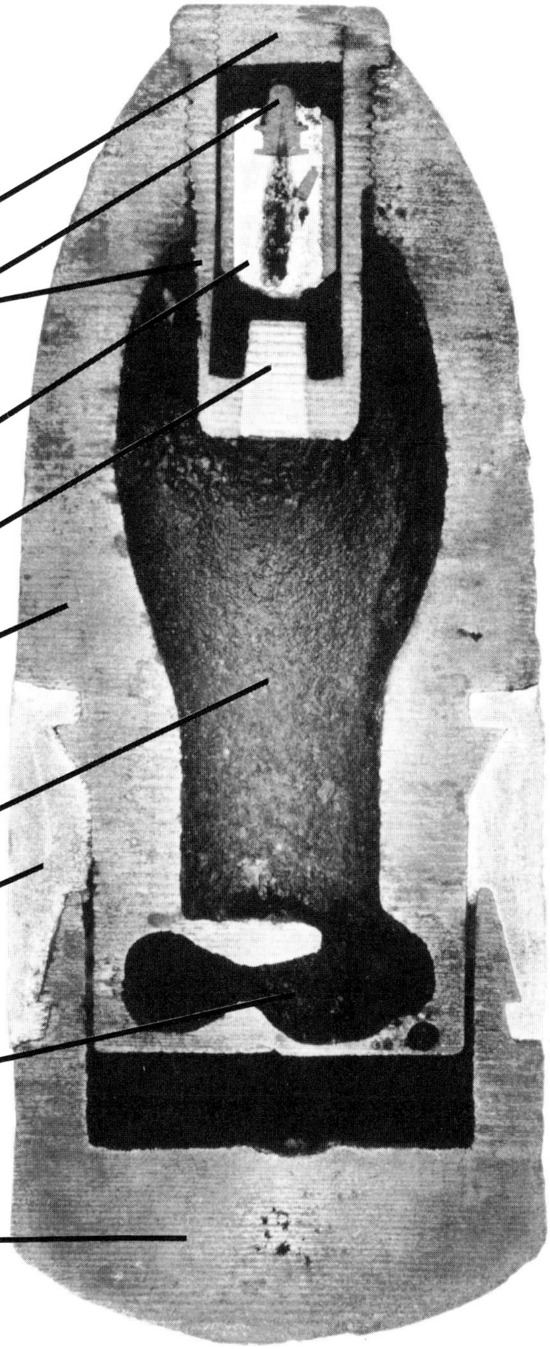

Comments: This Hotchkiss pattern was designed to explode at the time of impact. It did not require flame grooves on the projectile body due to the percussion fuse system. See page 131, figure III A-84, for a complete projectile. This is an example of a projectile that does not contain case shot. It is often referred to as a common shell.

Fig. II B-8

Federal

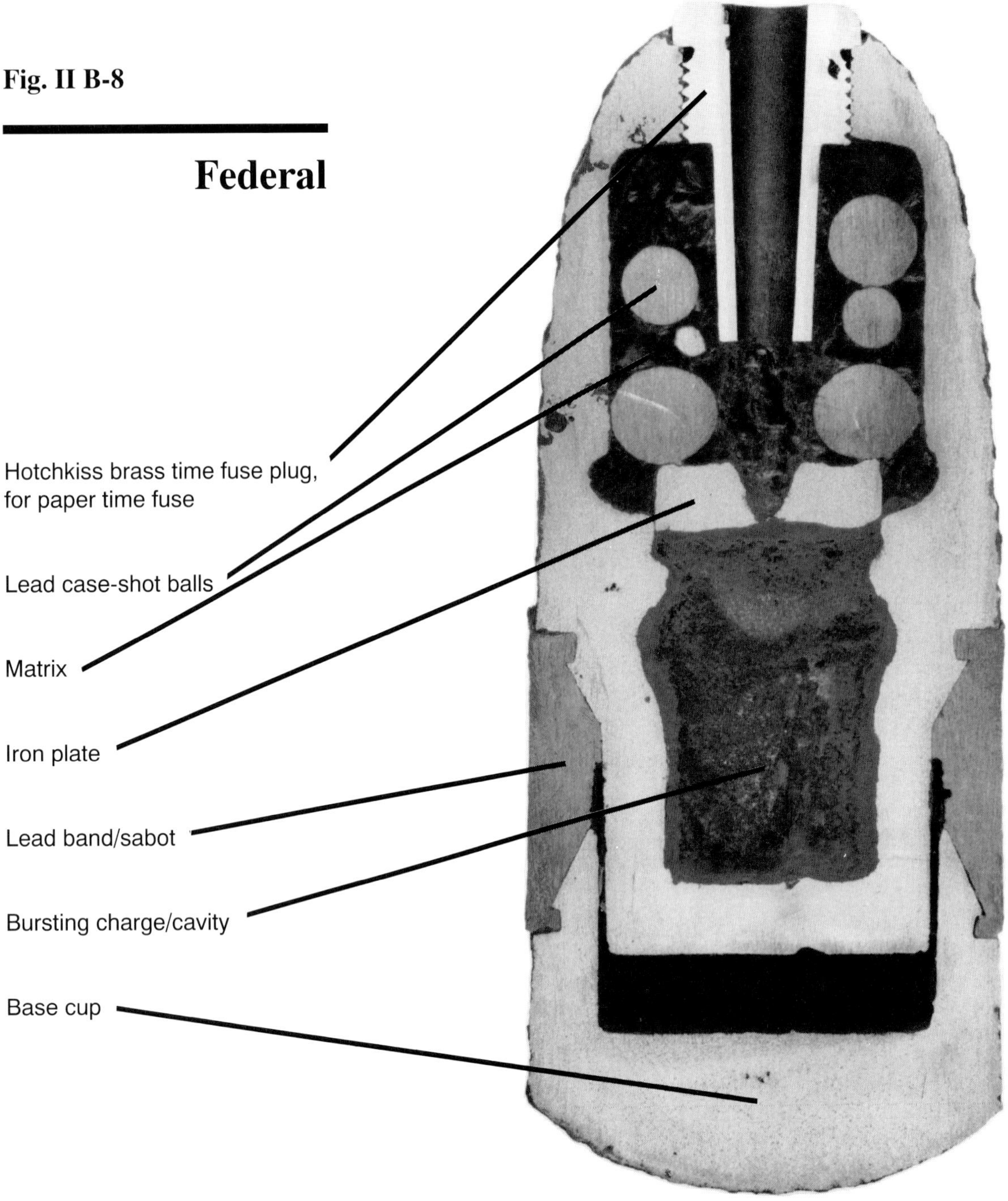

Comments: This is a case-shot variety of the Hotchkiss pattern. The iron plate improvement separated the internal bursting cavity, which aided in the detonation of the bursting charge. It was patented April 28, 1863, patent #38,359. The iron plate prevented the mixture of the case-shot balls with the bursting charge, thus causing the case-shot to be thrown forward with greater velocity. Iron case-shot balls have been found inside other specimens. See page 130, figure III A-83, for a complete specimen.

Fig. II B-9

Federal

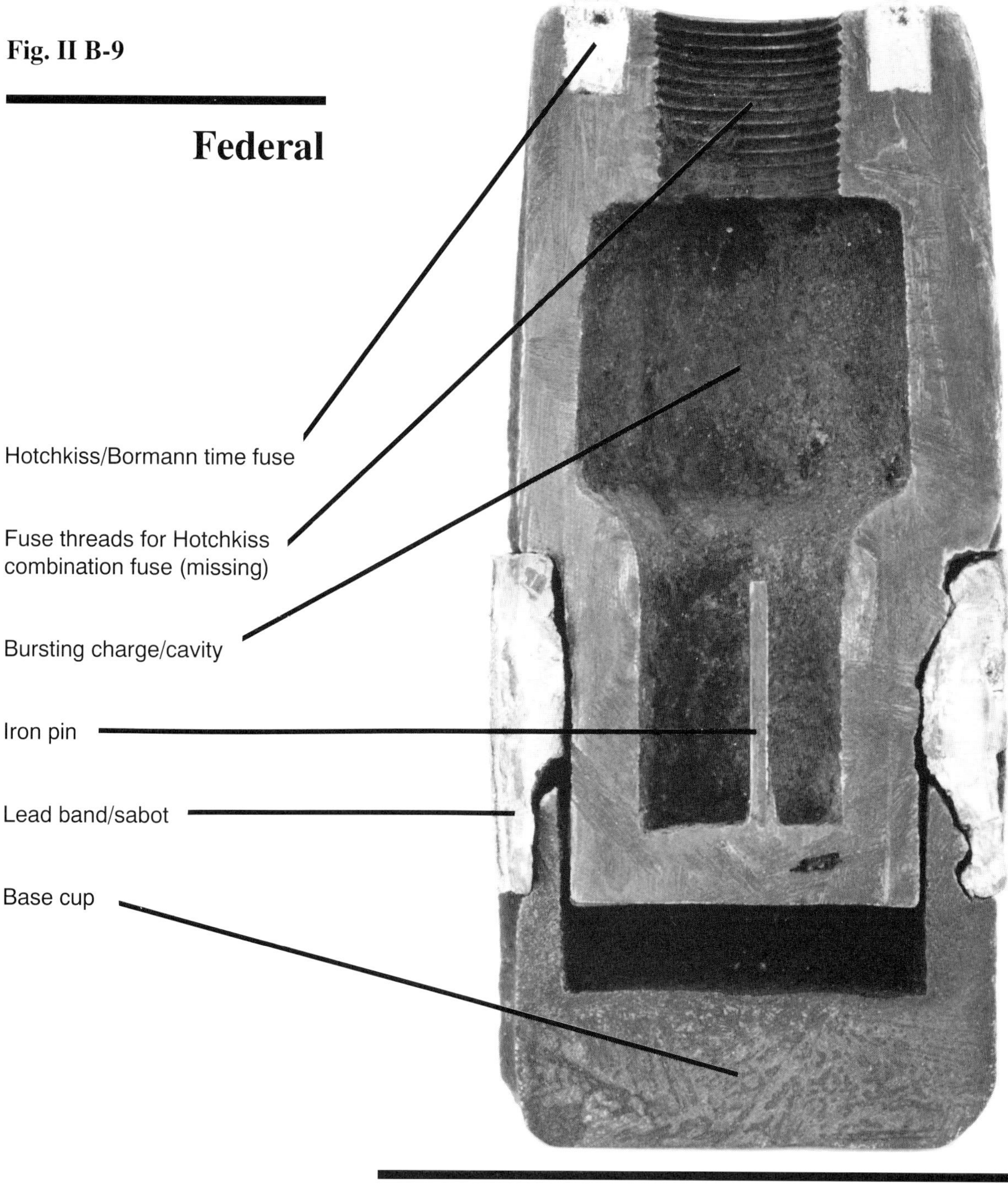

Comments: This Hotchkiss projectile is of post-war manufacture. The purpose of the iron pin in the base of the projectile's nose was to hold the internal sand casting core in place. See page 251, figure III B-14, for further comments and a complete specimen.

Fig. II B-10

Federal

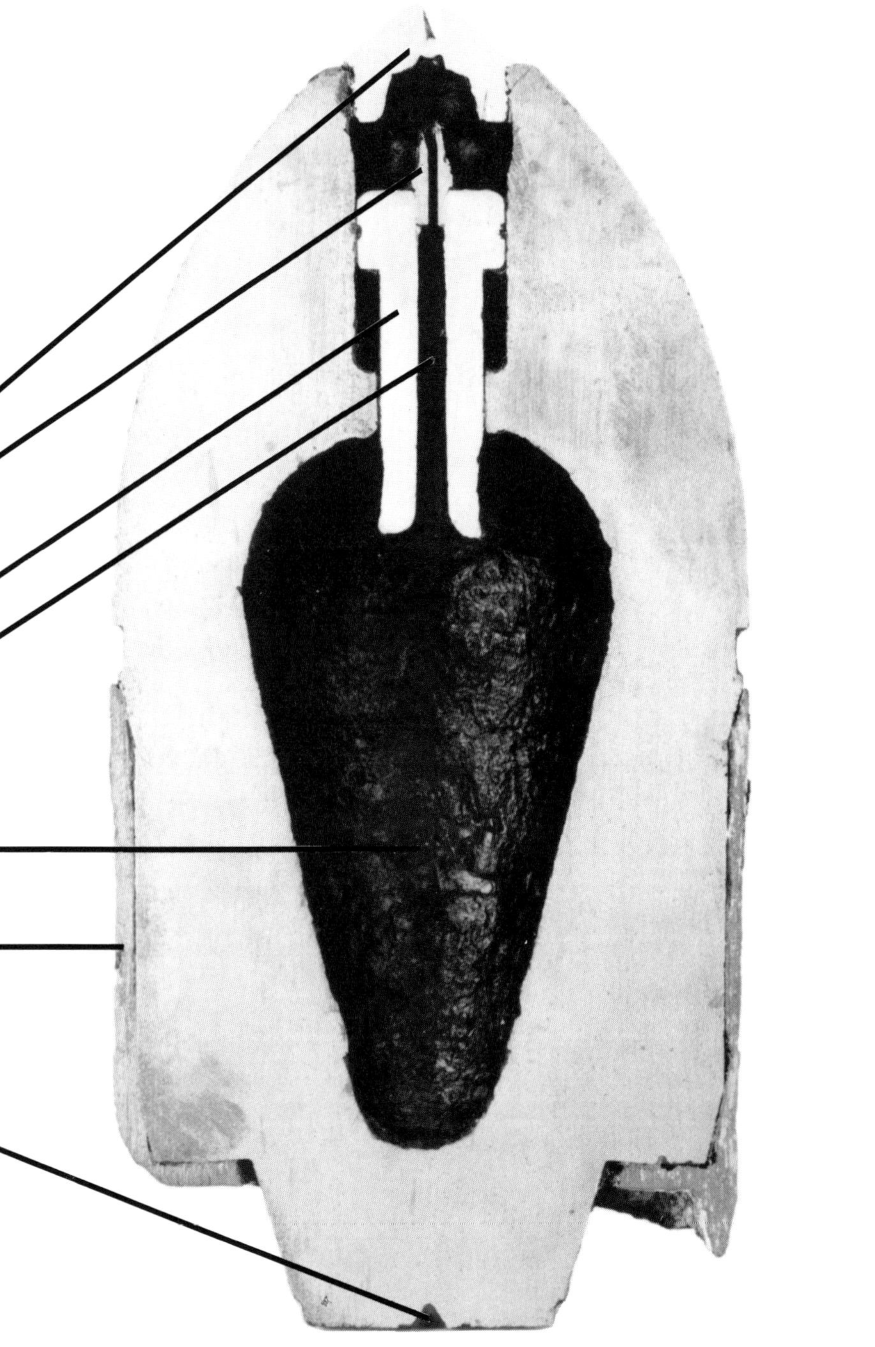

Comments: The pictured specimen is a James type II pattern projectile. The bursting cavity was too small, which resulted in poor fragmentation upon detonation. See page 150, figure III A-103, for additional comments and a complete specimen.

Fig. II B-11

Federal

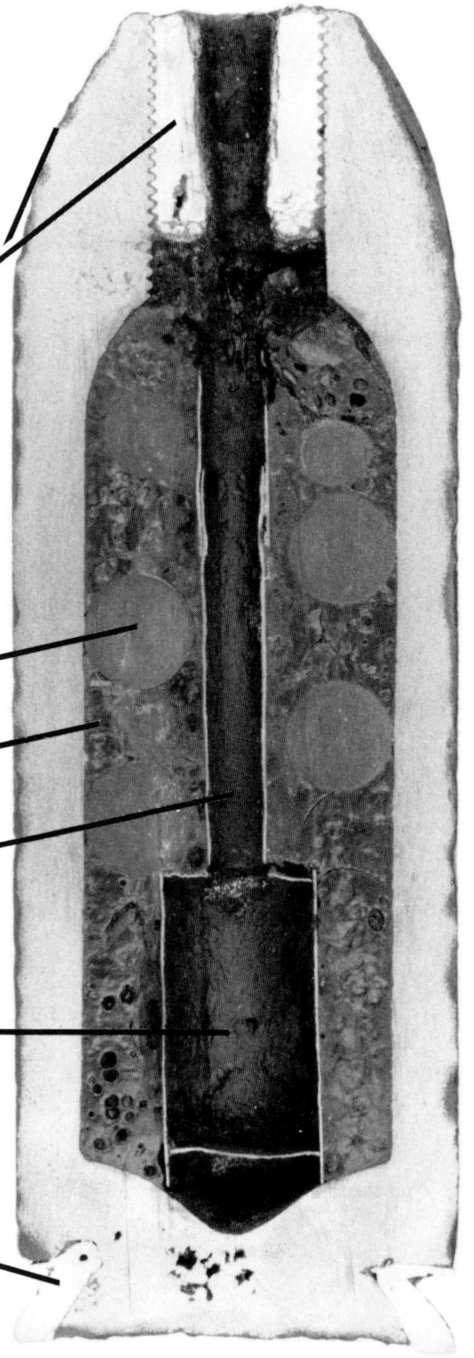

Comments: This is a good example of a Federal Parrott case-shot projectile that was manufactured in large quantities. The black pitch matrix was an improvement patented by Alfred Berney of New Jersey on June 17, 1862, patent #35,659. Mr. Berney used the residue from the distillation of coal-tar, generally known as "asphaltum," to create a matrix. The asphaltum replaced the sulphur matrix. See page 168, figure III A-121, for a complete specimen.

Fig. II B-12

Federal

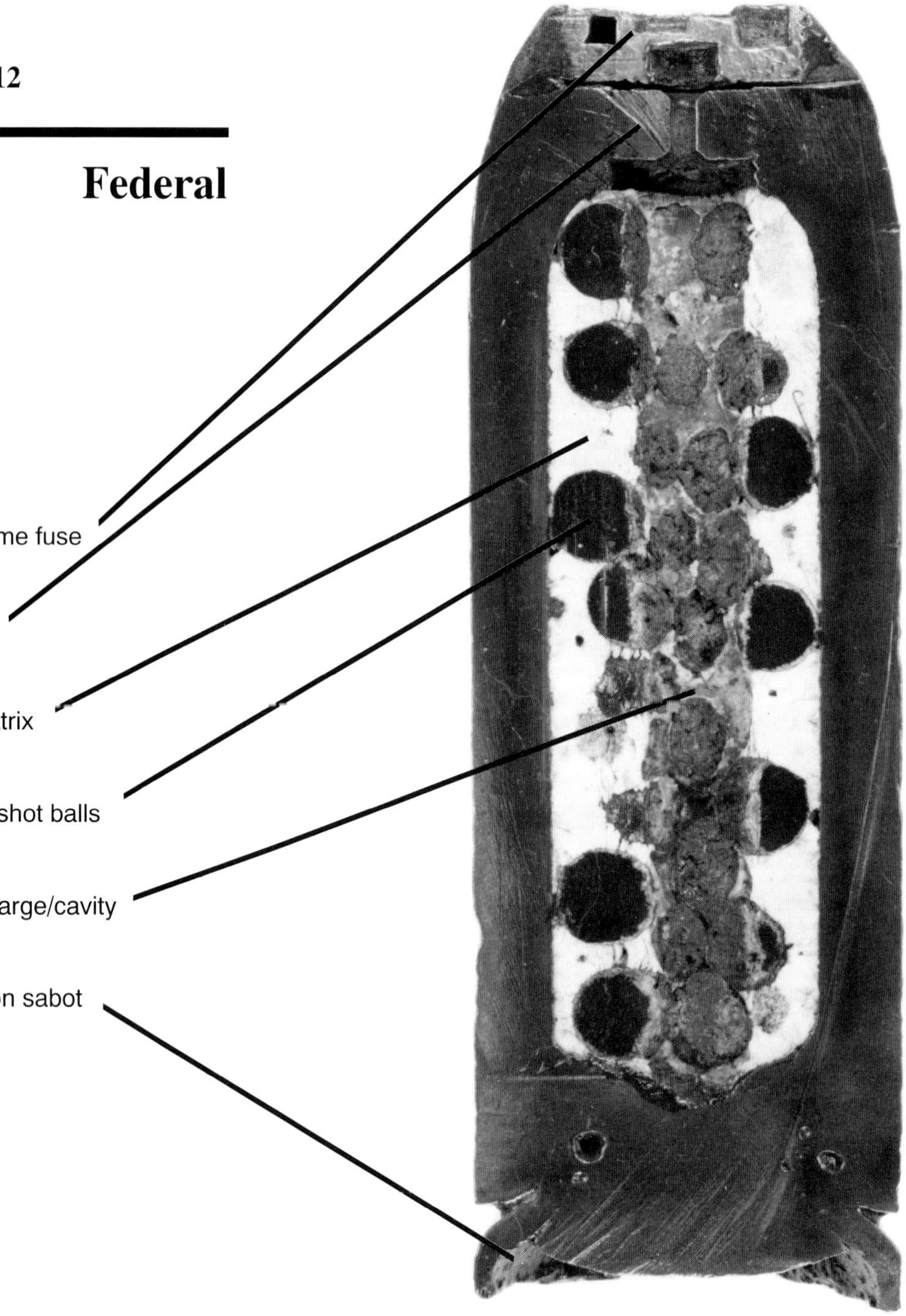

Comments: The sulphur matrix was replaced with a coal-tar distillate after 1862. Note that the bursting cavity was formed by drilling through the case-shot material. The Parrott projectile fitted with a Bormann time fuse is uncommon. The 5 1/4-second burning time of the standard Bormann time fuse was insufficient for rifled artillery. See page 166, figure III A-119, for further comments and a complete specimen.

Fig. II B-13

Confederate

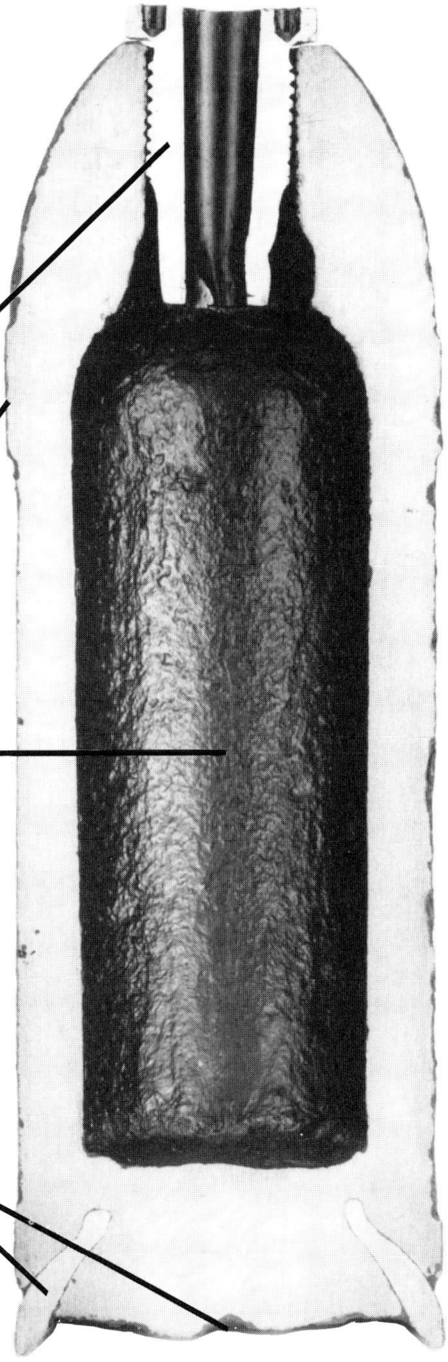

Comments: This projectile was manufactured by Tredegar Iron Works, in Richmond, Virginia, and is an example of a tall-pattern Read. See page 185, figure III A-138, for a complete specimen. This is an example of a projectile that does not contain case-shot balls. It is often referred to as a common shell.

Fig. II B-14

Confederate

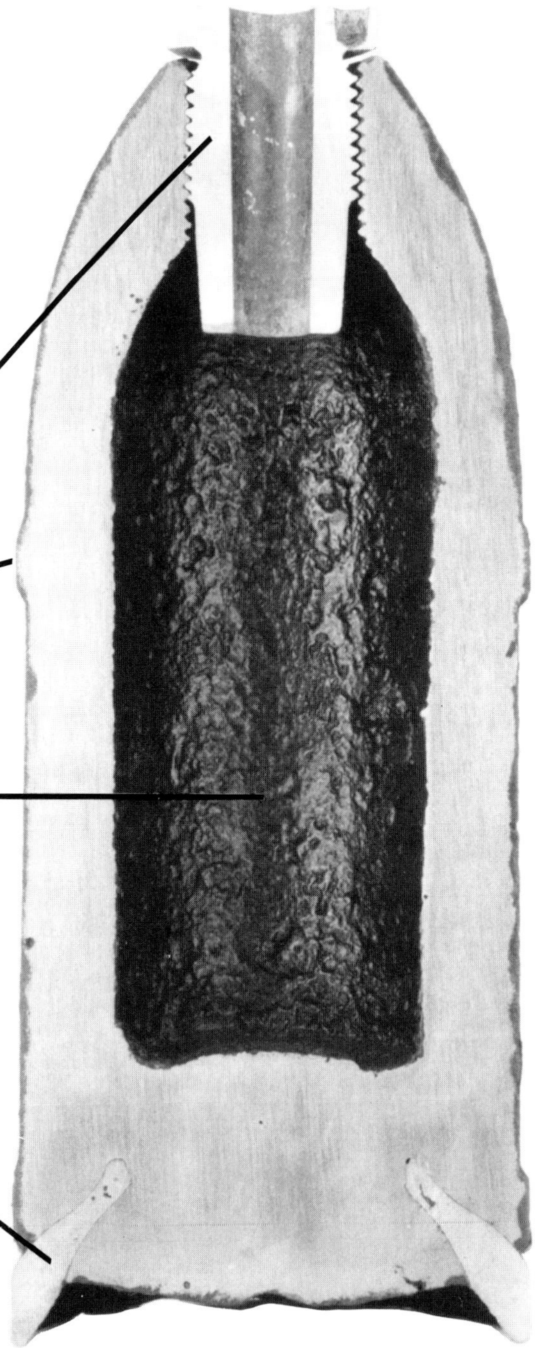

Comments: This is the classic style of bourrelet Confederate Read projectile. The projectile was manufactured in a two-step process: first the iron projectile body was cast, then the copper sabot was cast into the body of the projectile. This was due to the fact that copper has a lower melting point, rendering it impossible to cast the projectile in one step. See page 193, figure III A-146, for a complete specimen.

Fig. II B-15

Federal

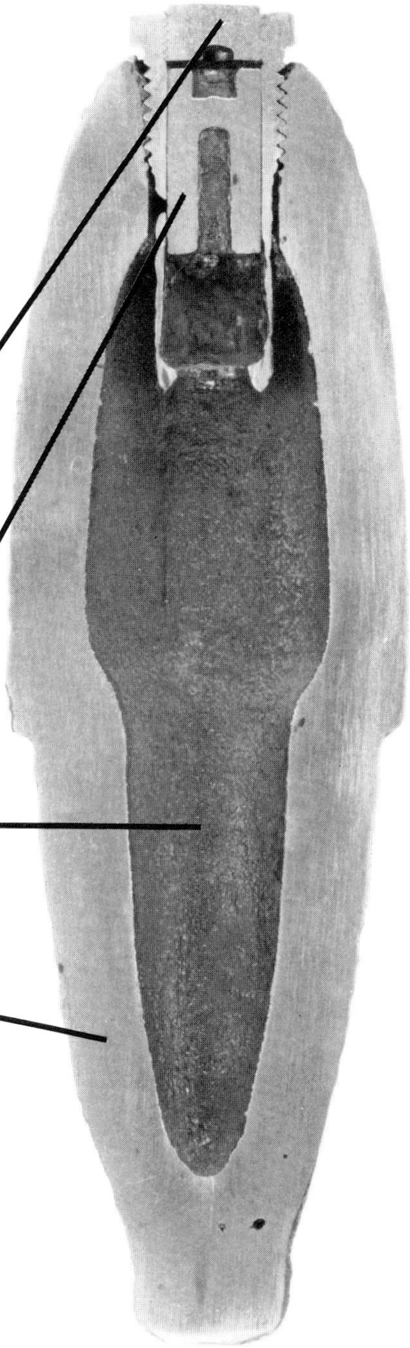

Schenkl brass percussion fuse:

Anvil cap

Plunger/striker

Bursting charge/cavity

Tapered cone

Comments: This is an example of the Federal 3-inch Schenkl projectile. This common shell has a bursting charge cavity that does not contain any case-shot material. See page 215, figure III A-168, for a complete specimen.

Fig. II B-16

Federal

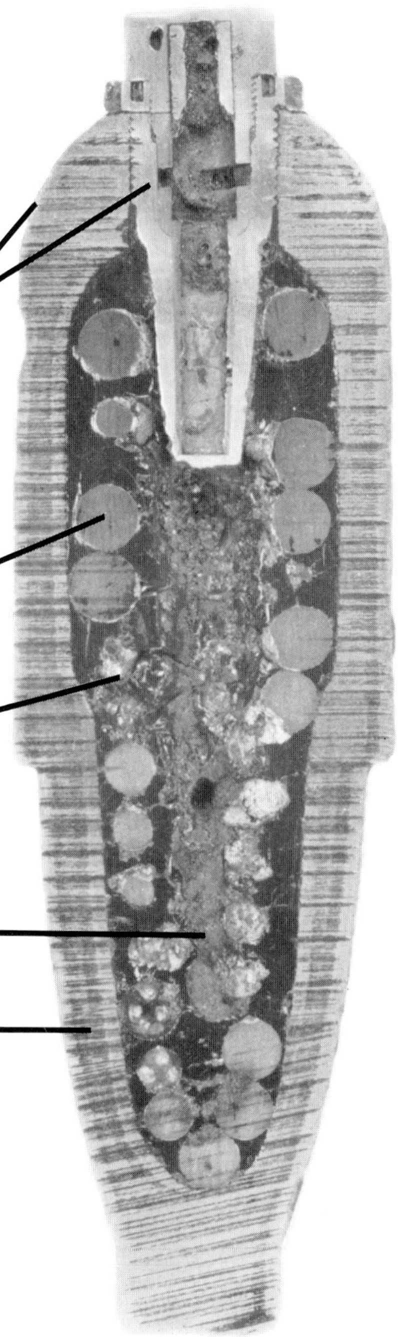

Comments: The more rounded nose on this Schenkl projectile indicates that it contains case shot. The Schenkl combination fuse was unreliable under combat conditions. See page 216, figure III A-169, for additional comments and a complete specimen. See page 42 for information on the fuse.

Fig. II B-17

Confederate

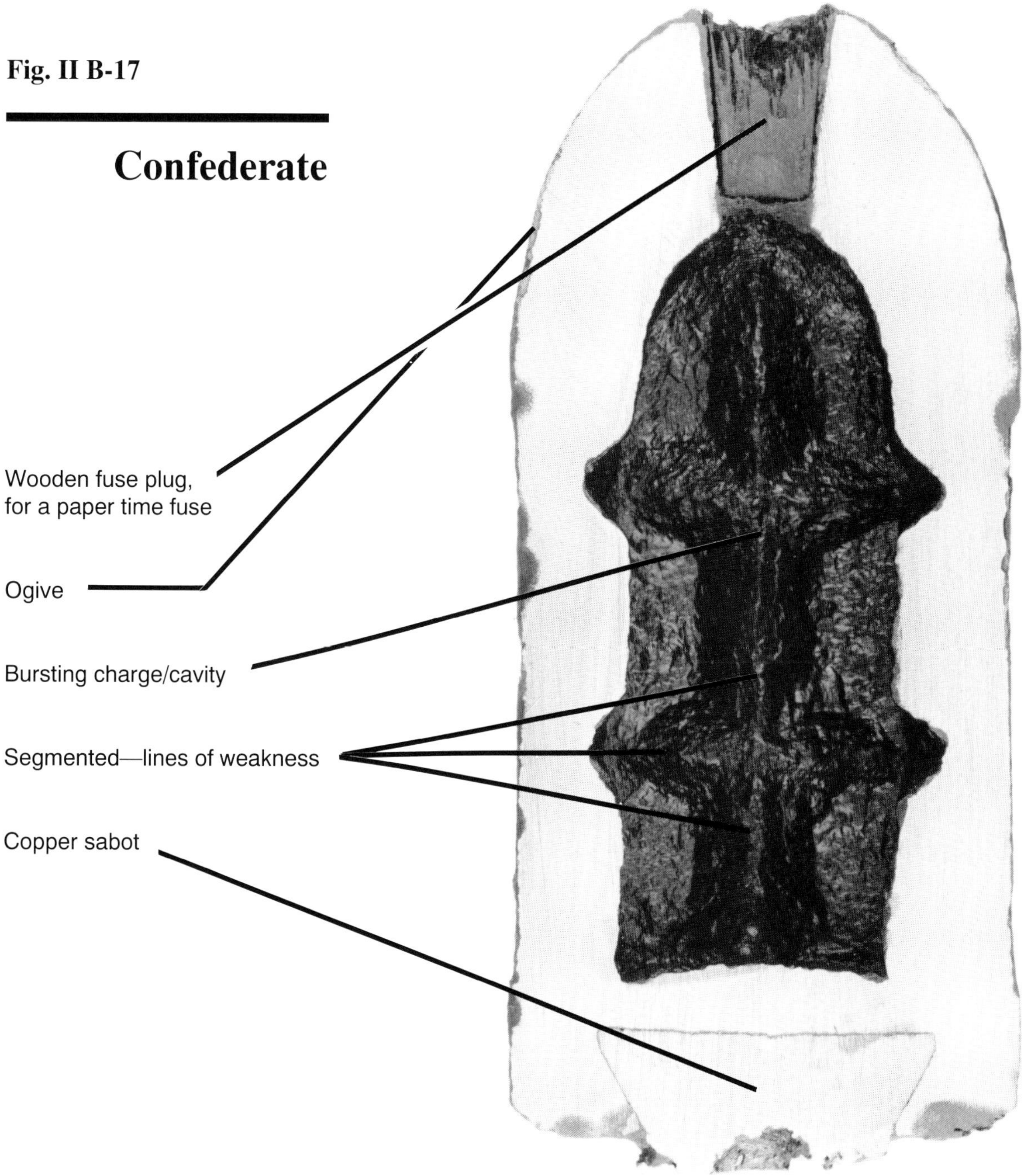

Comments: This is an example of a Selma Arsenal-manufactured segmented-interior projectile. Note that the copper sabot is partially missing. It was cast with the lines of weakness in the interior. See page 229, figure III A-182, for a complete specimen.

Fig. II B-18

Confederate

Comments: This is an example of a Confederate 12-pounder (4.62-inch caliber) spherical polygonal-cavity projectile. It was found exploded in the ground, recovered, and reassembled. The lines of weakness are clearly evident.

Fig. II B-19

Federal

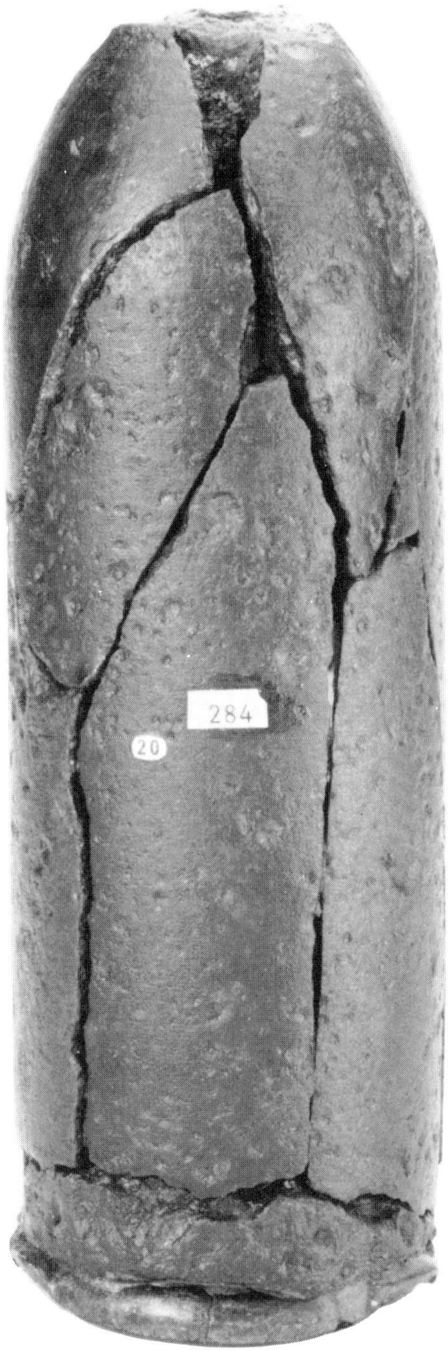

Comments: This is an example of a Federal 10-pounder (2.9-inch caliber) Parrott projectile. The projectile burst into random-size fragments. No two projectiles ever fragmented in a similar pattern unless they possessed a polygonal cavity or segmented interior.

Fig. II B-20

Federal & Confederate

Comments: This is an interior view of a standard 12-pounder (4.62-inch caliber) canister. Sawdust filled the spaces between the balls. The tin sides are missing, as is the wooden sabot. Note that the top plate is thinner than the bottom plate, which had to absorb the initial discharge force during firing. This pattern of construction is typical of both Federal and Confederate canisters. See page 76, figure III A-29, for a complete, non-excavated specimen.

II. Field Artillery Projectiles

C. Sabot and Stability Systems

A useful method of distinguishing artillery projectiles is by grouping them according to the sabot or stabilizing systems they employed. In order for the sabot or stabilizer to function properly, the artillery tube needed to be rifled in the familiar land and groove pattern. Some systems had their own distinctive pattern, such as the Whitworth cannon tube. Historically, nine different major stabilizing systems saw use during the Civil War.

There were nine different stabilizing systems employed during the course of the Civil War.

1 The most common system employed was developed by John B. Read. Closely resembling Read's system was the one used by Robert P. Parrott. In both variations when the cannon was fired, the force of the explosion caused a ring of soft metal attached to the base of the projectile expand into the spiral grooves of the tube. This imparted a rotary motion to the projectile. The expansion ring was commonly made of wrought iron, brass, copper, or lead. Projectiles using this system can be found on page 167, figure III A-120, and page 184, figure III A-137.

2 The second system consisted of a curved, or cupped plate, commonly made of brass, copper, or lead, secured to the projectile by a central screw or bolt. The plate was held firm either by a system of radial grooves in the iron projectile base or by three projections extending from the plate into holes in the base of the projectile. The plate could also have three or more iron pins protruding from the base of the projectile through holes in the plate.

The rifled motion was imparted to the projectile when the cannon detonation flattened the plate. This forced it into the grooves of the artillery tube by increasing its caliber.

Projectiles using this system type include the Mullane or Tennessee sabot (page 154, figure III A-107) and Brooke ratchet sabot (page 103, figure III A-56).

3 The third system employed a soft band of lead, as in the Hotchkiss projectile (page129, figure III A-82) or a papier-mâché cylinder, as in the Schenkl projectile (page 216, figure III A-169). These sabots were forced into the grooves of the cannon by expansion as they were pushed over a portion of the projectile body by the force of the propellant gases.

4 The fourth system was exemplified by the Britten and Dyer projectiles. A lead cup enclosed the projectile base and extended a varying distance up the body of the projectile. This sabot attachment method was patented by Bashley Britten, British patent #604, dated March 17, 1855. An example of a projectile using this system is on page 101, figure III A-54.

5 The fifth system was also of British origin. An example is the Armstrong projectile. This projectile had to be fired from a specially designed cannon tube. The system consisted of two or more rows of metal studs that projected from the body of the projectile. These studs fit into the deep side of the angled grooves of the cannon tube. When the firing charge was detonated, the projectile shifted to the less distinct end of each groove, which compacted the studs. This helped the projectile take the rifling during discharge.

6 The sixth system was similar to the Armstrong design but it used flanges instead of studs. The projectile had two or three long vertical iron projections that corresponded to the rifling of the cannon tube. The

Pattison projectile on page 173, figure III A-126, used this system.

7 The seventh system fitted a bore-shaped projectile with a cannon tube. The exterior of the projectile and the interior of the cannon bore corresponded with each other. The projectile was the exact diameter of the cannon bore, minus windage. Breechloading and muzzleloading cannon used this system. The Whitworth (page 230, figure III A-183) and Preston (page 174, figure III A-127) projectiles are examples of projectiles that used this system.

8 The eighth system is another of British design. It employed a breech-loaded projectile that had a diameter slightly larger than the bore of the cannon. The sabot was compressed into the grooves of the rifling upon discharge. The lead-covered Armstrong projectile on page 97, figure III A-50, is an example of a projectile that employed this system.

9 In the ninth system, the projectile body had raised and/or depressed areas on the base or nose, which were angled to cause rotation. This system was used in smoothbore cannon that fired conical projectiles. The Carkeet projectile on page 111, figure III A-64, is an example of a projectile that used this system. Another variation is noted in the wing shot.

II. Field Artillery Projectiles

D. How Fuses Worked

Civil War artillery fuses, just like the projectiles, came in many different sizes, shapes, and varieties. While the primary purpose of the fuse was to explode the projectile among enemy forces, the methods used to accomplish this objective varied. In many cases, the artillerist was the one who decided what type of fusing system to employ based on the range of the target, type of target, and type of projectile he intended to use. In other cases, the artillerist made the choice of projectiles and had to use the fusing system manufactured into the projectile.

All fusing systems can be broken down into four categories:

All of the fuzing systems can be broken down into four basic categories.

1 TIME

This system—the most employed fuse system—was designed to explode a projectile after a predetermined number of seconds. Two basic systems of time fuses were used. The first system was simply a powder composition, wrapped tightly in paper and pre-cut at the factory to determine the time before explosion. It was driven into a metal or wood fuse plug mounted in the fuse hole of the projectile. The second system employed a soft metal housing containing a powder train. The artillerist cut through the train at the appropriate time mark before loading the projectile into the cannon.

Both time fuse systems relied on the flame from the exploding powder charge in the cannon tube to ignite the powder composition. After the composition had burned down through the set number of seconds, the fuse flame would enter the powder chamber of the projectile and cause the projectile to explode.

The most common projectiles found that employed the time fuse system are Parrott, Hotchkiss, and Bormann-fused spherical shells.

2 PERCUSSION

Percussion was the second most employed fuse system. These were designed to explode the projectile when the fuse made contact with an object. Most percussion fuses employed a plunger-and-anvil method of detonation. The fuse body was usually brass, copper, or alloy, and was threaded to be screwed into the nose of the projectile. When the fuse struck an object, a striker was driven down a chamber inside the fuse and struck a percussion cap (which was seated on a nipple) against the anvil. Employing the same principle as a musket, the percussion cap then exploded, sending a small flame through a hole in the nipple and igniting a powder train. The powder train transferred the flame to the powder chamber and detonated the projectile.

The common projectiles that employed the percussion fuse system are Parrott, Hotchkiss, James, and Schenkl.

3 COMBINATION

As the name implies, this system was a combination of the time fuse and the percussion fuse systems. Combination fuses varied widely in their method of detonation depending upon the inventor and manufacturer. The basic premise was that the fuse was set prior to firing to a specific detonation time. When the projectile was fired from the cannon tube, the shock (or inertia) of the firing caused a plunger to slide down and strike a chemical composition against a metal platform, causing a flame to ignite the powder train. If the projectile struck an object before the time element was completed, the system was designed to activate as a percussion fuse.

Combination fuse systems can be found on Armstrong, Schenkl, and Sawyer projectiles, among others.

4 CONCUSSION

These were designed to activate from the shock of striking an object. Once again, the actual mechanics of the fuse system varied according to the manufacturer and inventor. Some systems employed chemical vials which, when shattered upon impact, caused a fire inside the fuse body and transferred to the powder chamber. Other systems relied upon plaster to separate the igniting material from the powder until the plaster shattered upon impact.

Warning: Chemical concussion fuses should be avoided by collectors unless it is known beyond any doubt that the vials have already been destroyed.

Many projectiles found in the field still retain their fusing system. Reasons for the failure of a fuse system to detonate are many, but the most common explanations are that the fuse was damaged while being set into the projectile (by either the manufacturer or the artillerist), the fuse was damaged upon impact, or, as in the case of a percussion system, the projectile tumbled as it traveled and landed without the fuse striking anything. Many projectiles recovered from battlefields indicate that the fuse worked properly but the projectile was defective.

The illustrations that follow are not meant to be inclusive or comprehensive with respect to all classes and examples, but rather to give the reader an overview of this fascinating area of study.

Bormann Time Fuse

Federal 5 1/4-second Bormann time fuse

Confederate 5 1/2-second Bormann time fuse (copy)

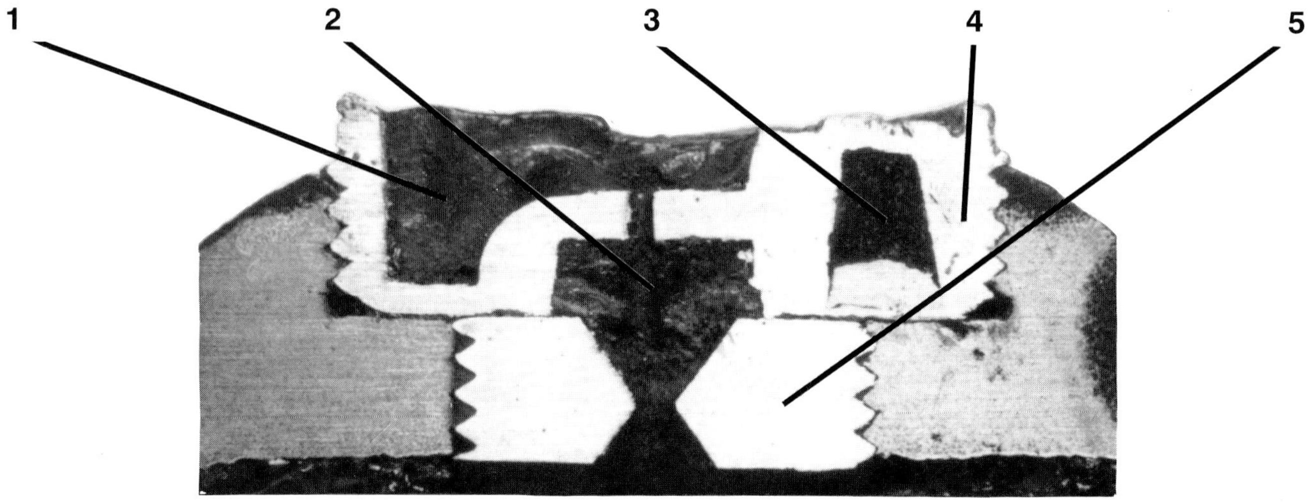

The Bormann is named after its inventor, Belgian Army Captain Charles G. Bormann. The time fuse is contained in a lead and tin disk (4). This disk has time markings indicated in seconds and quarter seconds graduated up to about 5 1/4 seconds. The artillerist used a metal punch to pierce the thin metal at the desired time marking. This exposed a horseshoe-shaped horizontal powder train (3), which was covered by a thin sheet of tin. When the cannon discharged, the flame from the explosion ignited this powder train. It burned at a uniform rate in both directions, but one end terminated at a dead end just beyond the 5 1/4-second mark (some specimens are 5 1/2 seconds). The other end continued to burn past the zero mark, where it traveled through a channel (1) to a small powder booster or magazine (2). This powder then exploded, sending the flame through a hole in the fuse underplug (5) to the powder chamber of the projectile. The purpose of the brass or iron fuse underplug was to form a solid base for the fuse which, because of its soft metal, could have easily been damaged during firing.

Hotchkiss Time Fuse

This Hotchkiss fuse plug is made of brass and has a screwdriver slot across the face. The plug is threaded and was screwed into the fuse hole of the projectile. A paper time fuse, composed of a powder composition tightly wrapped in paper, was then driven into the center of the plug. The length and composition of the fuse determined the amount of time it took to finish burning. The artillerist could either choose a manufactured fuse or alter the length of a fuse so that it would burn for a shorter period of time. When the fuse burned through to the end, the flame was transmitted through the hole in the fuse plug and into the powder chamber of the projectile. This is a representative example of this class of time fuses.

Schenkl Combination Fuse

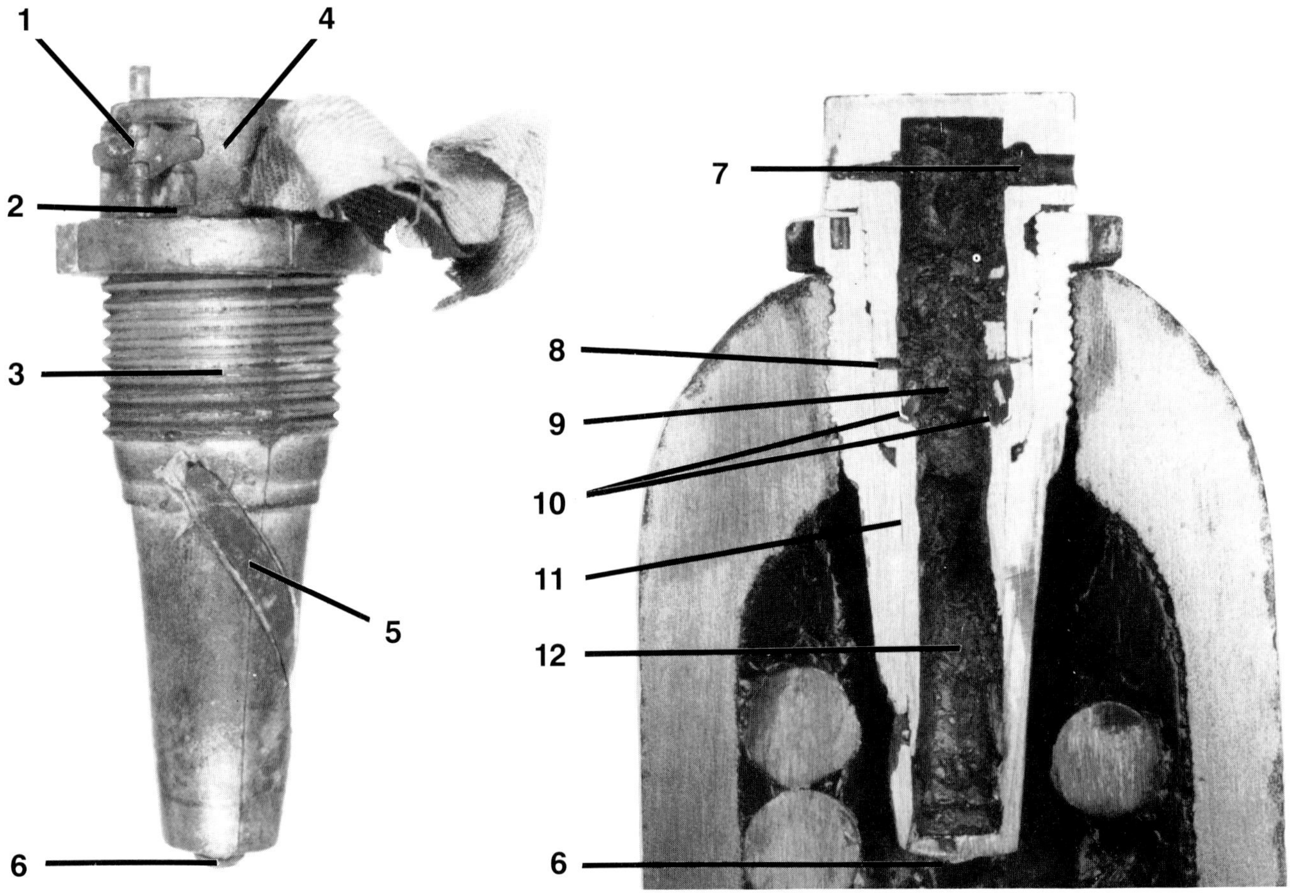

This fuse stock (3) is made of zinc and has a time index (2) on the face of the top flange graduated from zero to 9 1/2 seconds. The fuse body is closed at the bottom (6) and pierced by a series of small holes (5) spiraling down from the middle of the fuse body to the end. The rest of the fuse consisted of a combustible fuse compound (12) contained in a metal case that also has holes spiralling down but in an opposite direction from the body. A rotator (4) is screwed into the top of the fuse. Inside the rotator is the plunger, which is held in place by a soft metal pin and an iron safety pin. An interior fuse rotator (11), affixed to the top rotator by two metal dowels, contains a fulminate ring (10) and a highly combustible soluble cotton compound (9). When the artillerist was ready to arm the fuse, he removed the safety pin (placed where 7 points; pin is missing from this cutaway) and lifted the spring clip (1), set the rotator to the correct time (using a stamped arrow as a guide), and then locked the rotator in place. By doing this the two sets of holes lined up with each other to produce the proper burning time. The inertia of the cannon firing caused the internal safety pin (placed where 8 points; pin is missing from this cutaway) to break and threw the plunger back against the ring. The flame from that action caused the cotton to flare up and lighted the time fuse compound. The flame burned through the two holes into the projectile's powder chamber. If the projectile struck anything before the time fuse was through burning, the fuse body would break, causing the burning compound to drop immediately into the powder chamber.

Parrott Percussion Fuse

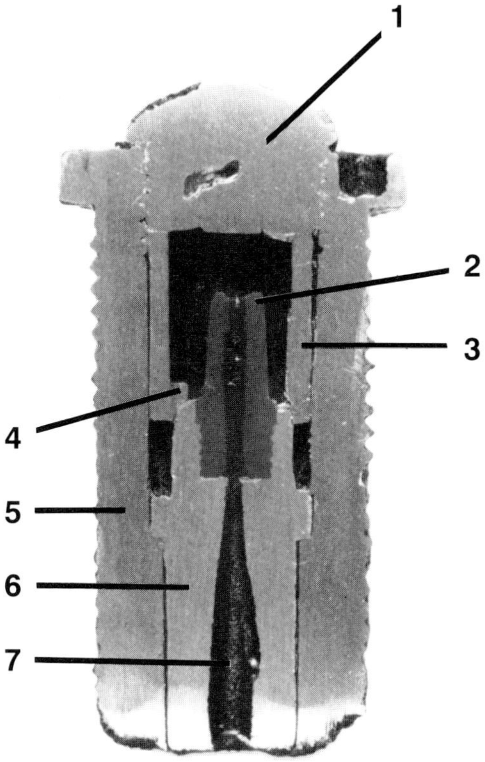

This type of Parrott percussion fuse saw extensive use. The fuse body (5) is made of zinc and was screwed into the fuse hole by a spanner wrench using the two spanner holes at the top of the fuse. An anvil (1) is screwed into the top center of the fuse. The interior contains a percussion cap (placed where 2 points; missing from this cutaway) seated on a nipple, the plunger or striker (6), and the powder train (7). The striker is held back by two small projections (4) on a safety pin (3). When the fuse impacted, the two projections broke and allowed the striker to move forward. The percussion cap struck the anvil and exploded, sending a flame through the nipple to the powder train. The flame was then transferred to the powder chamber of the projectile. This is a representative example of this class of percussion fuses.

II. Field Artillery Projectiles

E. Quick Identification Of Fuses

On the following two pages are numbered examples of selected artillery projectile fuses. The numbers correspond to the fuses shown on the next two pages. We have also made reference to representative projectiles that employed these fuses.

(1) Packet of five paper time fuses from Frankford Arsenal
(2) Packet of five paper time fuses—10-second time fuses
(3) Eureka time fuse (page 249, figure III B-12)
(4) Absterdam percussion fuse
(5) Britten percussion fuse (page102, figure III A-55)
(6) Britten percussion fuse for Preston (page 174, figure III A-127)
(7) Parrott percussion fuse (page 165, figure III A-118)
(8) Parrott percussion fuse
(9) Schenkl percussion fuse (page 215, figure III A-168)
(10) Schenkl combination fuse (page 216, figure III A-169)
(11) Sawyer percussion fuse (page 213, figure III A-166)
(12) Sawyer combination fuse (page 212, figure III A-165)
(13) Confederate West Point Type percussion fuse (page 209, figure III A-162)
(14) James percussion fuse (page 150, figure III A-103)
(15) Hotchkiss percussion fuse (page 136, figure III A-89)
(16) Paper time fuse
(17) Confederate Bormann (copy) time fuse (page 50, figure III A-3)
(18) Underplug for Bormann time fuse (page 18, figure II B-4)
(19) Federal Bormann time fuse (page 53, figure III A-6)
(20) Britten time fuse
(21) Absterdam time fuse (page 240, figure III B-3)
(22) Taylor time fuse (page 247, figure III B-10)
(23) Confederate replacement for the Bormann time fuse (page 61, figure III A-14)
(24) Confederate spherical projectile time fuse (page 60, figure III A-13)
(25) Confederate conical projectile time fuse (page 184, figure III A-137)
(26) Confederate conical case-shot projectile time fuse (page 188, figure III A-141)
(27) Brooke time fuse (page 105, figure III A-58)
(28) Wooden time fuse plug for rifled projectiles (page 88, figure III A-41)
(29) Wooden time fuse plug for spherical projectiles (page 57, figure III A-10)
(30) Confederate screwdriver slotted time fuse for spherical projectiles
(31) Hotchkiss time fuse, brass (page 129, figure III A-82)
(32) Hotchkiss time fuse for case-shot, brass (page 130, figure III A-83)
(33) Hotchkiss time fuse for case-shot, lead (page 127, figure III A-80)
(34) Hotchkiss time fuse for case-shot, brass
(35) Parrott time fuse
(36) Parrott time fuse (page 168, figure III A-121)
(37) Parrott time fuse (page 171, figure III A-124)

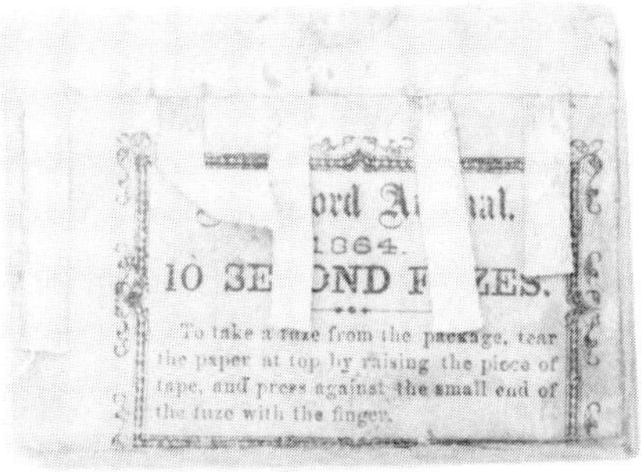

1

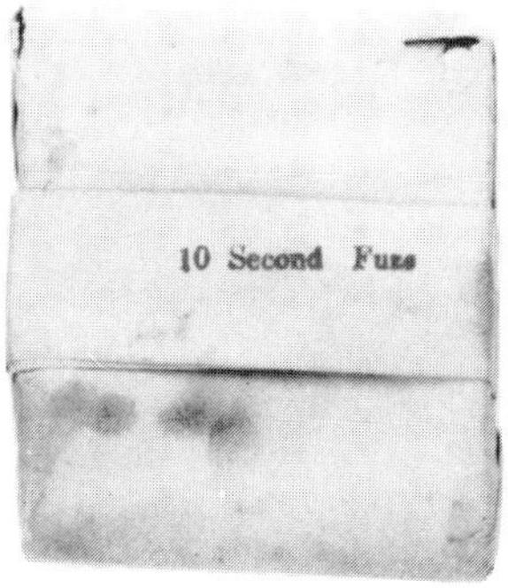

2

3

4

5

6

7

8

9

10

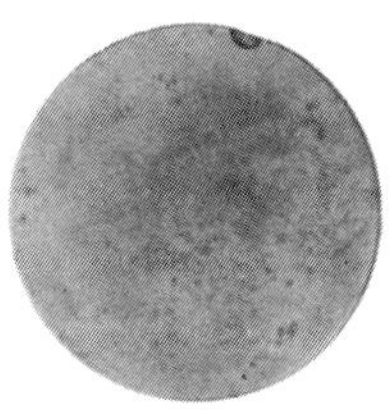

11

12

13

14

15

16

17

19

20

21

22

18

23

24

25

26

27

28

29

30

31

32

33

34

35

36

37

III. A Pictorial Study

A. Combat Proven

How to use this section

The following pictorial guide has been designed to be of maximum benefit to the beginning collector as well as the more advanced student. Each page is designed to serve as a quick reference guide for the most common varieties of artillery projectiles. With careful review, the reader will find valuable information to help point out many of the more subtle differences between apparently similar projectiles.

A note to the reader regarding the sizes of the projectiles shown: No attempt was made to show the projectiles at their actual size. It would have been impossible to do so with many of the larger projectiles; further, many of the smaller projectiles would have been rendered too small for the reader to clearly see their special characteristics.

As an alternative, we have elected to show the *relative* gradations of size. The larger shells will appear so on the page; the smaller specimens will appear slightly smaller, but will still be large enough to make out important details. In every case, the reader should refer to the length and diameter measurements in the left column for actual size.

The comments at the bottom of the page deserve special attention. In addition to providing information concerning the shell illustrated, they may also refer the reader to other sections of the book to help clarify a point or to expand on a particular thought.

In an attempt to keep collectors from getting confused when reading other reference material on projectiles, the authors have decided to continue referring to various patterns by their currently accepted common name. However, based on our research serious questions have developed concerning how certain pattern projectiles should be grouped. Examples are Brouns, Mullanes (Tennessee-sabot shells), Parrotts, Reads, and Selmas. Should the examples listed be referred to as Read/Brouns, Read/Mullanes, Read/Parrotts, Selma Arsenal-manufactured Selma/Reads, etc.? The authors hope that at some point in the future serious students of Civil War artillery projectiles will be able to form a consensus of opinion and sort out the answers.

Fig. III A-1

Federal

PATTERN: Spherical
DIAMETER: 3.57 inches
GUN: 6-pounder smoothbore, 3.67-inch caliber
LENGTH: —
WEIGHT: 6 pounds, 5 ounces
TYPE: Solid shot
SABOT: Wooden cup (missing)
FUSING: None

Comments: This is the most common 6-pounder projectile used by both Federal and Confederate forces. According to Gibbon's *The Artillerist's Manual,* " . . . a 6- or 12-pound ball will go through six men at 800 yards' distance." Gibbon also states that ". . . solid shot is of very little effect upon earthen parapets, unless it passes through, in which case it soon knocks them down. At 600 yards, shot from field-guns will penetrate from 5 to 6 1/2 feet into newly thrown-up earth."

Fig. III A-2

Confederate

PATTERN:	Spherical
DIAMETER:	3.55 inches
GUN:	6-pounder smoothbore, 3.67-inch caliber
LENGTH:	4 1/8 including sabot
WEIGHT:	3 pounds, 6 ounces
TYPE:	Shell
SABOT:	Wooden cup
FUSING:	Wooden fuse plug, paper time fuse

Comments: This non-excavated specimen rests on its original wooden sabot. The ball is affixed to the sabot by a ring or collar around the fuse opening and secured to the sabot with four iron straps nailed to the wooden cup. Note the crude mould seam around the ball, typical of Confederate-manufactured projectiles. Note also the deep groove at the base of the wooden sabot, which was used to help tie the cartridge bag in place.

Fig. III A-3

Confederate

PATTERN:	Spherical
DIAMETER:	3.58 inches
GUN:	6-pounder smoothbore, 3.67-inch caliber
LENGTH:	—
WEIGHT:	4 pounds, 8 ounces
TYPE:	Shell
SABOT:	Wooden cup (missing)
FUSING:	Confederate Bormann time fuse

Comments: This is the Confederate copy of the Federal Bormann fuse and projectile. The Federal artillery used this pattern until the 6-pounder smoothbore cannon was removed from the field in 1864. One indicator of Confederate manufacture is that the starting line on the fuse is full length. The Federal Bormann fuse has a starting line that is half of the length of the Confederate Bormann fuse. Another indicator is that the Federal Bormann fuse was only 5 1/4 seconds. The 5 1/2-second Bormann fuse was discontinued by the Confederates by 1863 due to difficulty in manufacturing and poor battlefield performance. See page 40 for a photograph of the Bormann fuse.

Fig. III A-4

Federal

PATTERN: Spherical
DIAMETER: 4.50 inches
GUN: 12-pounder smoothbore, 4.62-inch caliber
LENGTH: 5 1/8 inches including sabot
WEIGHT: 12 pounds, 5 ounces
TYPE: Solid shot
SABOT: Wooden cup
FUSING: None

Comments: This specimen is non-excavated and rests on its original wooden sabot. This is the most common solid-shot projectile used during the Civil War by both the Federal and Confederate forces. The ball is secured to the sabot by two crossed iron straps nailed to the wooden cup. The cartridge bag or powder bag was tied to the groove cut into the base of the sabot. After attachment of the powder charge, the round was referred to as fixed ammunition.

Fig. III A-5

Federal

PATTERN:	Spherical
DIAMETER:	4.49 inches
GUN:	12-pounder smoothbore, 4.62-inch caliber
LENGTH:	—
WEIGHT:	11 pounds, 13 ounces
TYPE:	Solid shot
SABOT:	Wooden cup (missing)
FUSING:	None

Comments: This was the standard smoothbore projectile used during the Civil War. Its purpose was to destroy fortifications, cannon emplacements, trenches, and buildings. When fired on hard, dry ground it often ricocheted from one target to another. It is also found in the following calibers: 6-pounder (3.67-inch), 18-pounder (5.3-inch), and 24-pounder (5.82-inch) solid shot.

Fig. III A-6

Federal

PATTERN: Spherical
DIAMETER: 4.52 inches including straps
GUN: 12-pounder smoothbore, 4.62-inch caliber
LENGTH: 4 15/16 inches including sabot
WEIGHT: 8 pounds, 7 ounces (if case shot, 11-12 pounds)
TYPE: Shell
SABOT: Wooden cup
FUSING: Bormann time fuse

Comments: This specimen is non-excavated. The wooden cup was designed to align the fuse and keep it centered in the cannon bore. The cartridge bag was tied in the deep groove cut into the base of the sabot. After attachment of the cartridge bag, the round was referred to as fixed ammunition. The twine was used to tighten down the four tin straps, holding the sabot and fuse collar more securely. It was the most common spherical Federal shell used with the Bormann fuse in 12-pounder (4.62-inch caliber). It is found in the following field calibers with the Bormann time fuse: 6-pounder (3.67-inch) and 24-pounder (5.82-inch).

Fig. III A-7

Federal

PATTERN:	Spherical
DIAMETER:	4.52 inches
GUN:	12-pounder smoothbore, 4.62-inch caliber
LENGTH:	—
WEIGHT:	9 pounds, 4 ounces
TYPE:	Shell
SABOT:	Wooden cup (missing)
FUSING:	Bormann time fuse

Comments: The 12-pounder Bormann-fused projectile was used extensively by both Federal and Confederate forces. Note the star, which is an arsenal identification mark. The projectile is non-excavated and the Bormann fuse is unpunched. The fuse is made of an alloy of lead and tin. The function of the Bormann fuse is explained on page 40. A cross-section of a case-shot Bormann-fused projectile can be found on page 16, figure II B-2.

Fig. III A-8

Confederate

PATTERN:	Spherical
DIAMETER:	4.52 inches
GUN:	12-pounder smoothbore, 4.62-inch caliber
LENGTH:	—
WEIGHT:	11 pounds, 1 ounce
TYPE:	Case shot
SABOT:	Wooden cup (missing)
FUSING:	Confederate copy of the Bormann time fuse

Comments: Due to the repeated poor performance of the Confederate Bormann fuse copy, its manufacture was discontinued by 1863. Remaining Bormann fuse projectiles were drilled in the center of the fuse and a paper time fuse was inserted in its place.

Fig. III A-9

Federal

PATTERN:	Spherical
DIAMETER:	4.49 inches
GUN:	12-pounder smoothbore, 4.62-inch caliber
LENGTH:	—
WEIGHT:	8 pounds
TYPE:	Shell
SABOT:	Wooden cup (missing)
FUSING:	Bormann Braille time fuse

Comments: This unpunched Bormann Braille time fuse has dots instead of numbers. The dots are grouped, from one to five, in the location where the numbers would be on a standard-style Bormann fuse. No Confederate copies of this style of fuse have been recovered.

Fig. III A-10

Confederate

PATTERN:	Spherical
DIAMETER:	4.49 inches
GUN:	12-pounder smoothbore, 4.62-inch caliber
LENGTH:	—
WEIGHT:	9 pounds, 7 ounces
TYPE:	Shell, polygonal cavity
SABOT:	Wooden cup (missing)
FUSING:	Wooden fuse plug, paper time fuse

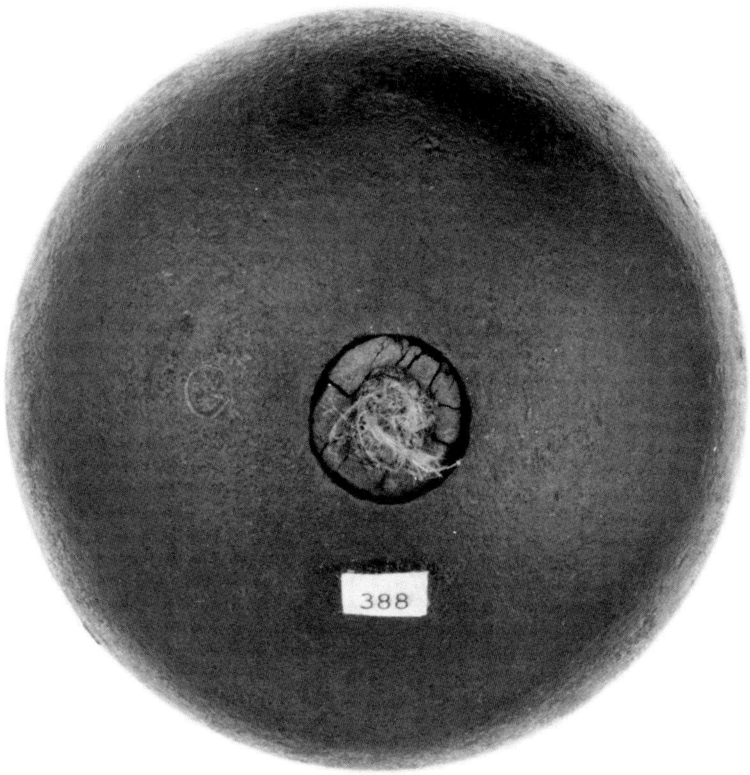

Comments: This shell is commonly called a polygonal-cavity projectile and was used exclusively by the Confederates. The interior was cast with a weakness in the form of pentagonal sections so that it would fragment into twelve equal pieces. It is also found in the "diamond" and "trapezoid" patterns. Note the "G" stamped to the left of the fuse plug, which is attributed to manufacture at Selma Arsenal in Alabama. The wooden fuse plug still has the packing toe inserted in it. The packing toe protected the powder charge from any foreign matter and was removed prior to the insertion of the paper time fuse. This type of shell is also found in 6-pounder (3.67-inch), and 24-pounder (5.82-inch caliber) smoothbore projectiles. See page 19, figure II B-5, for a cross-section.

Fig. III A-11

Confederate

PATTERN:	Spherical
DIAMETER:	4.47 inches
GUN:	12-pounder Coehorn mortar, 4.62-inch caliber
LENGTH:	—
WEIGHT:	7 pounds, 10 ounces
TYPE:	Shell
SABOT:	None
FUSING:	Wooden fuse plug, paper time fuse

Comments: The indentations, or ears, located on either side of the fuse hole were designed to aid in loading and aligning the projectile so that the fuse was in the center of the mortar's bore, just opposite from the powder charge. According to Abbot's *Siege Artillery,* the Confederate use of ears was far better than the clumsy Federal system of strapping with tin. See page 63, figure III A-16, for an example of the Federal projectile.

Fig. III A-12

Confederate

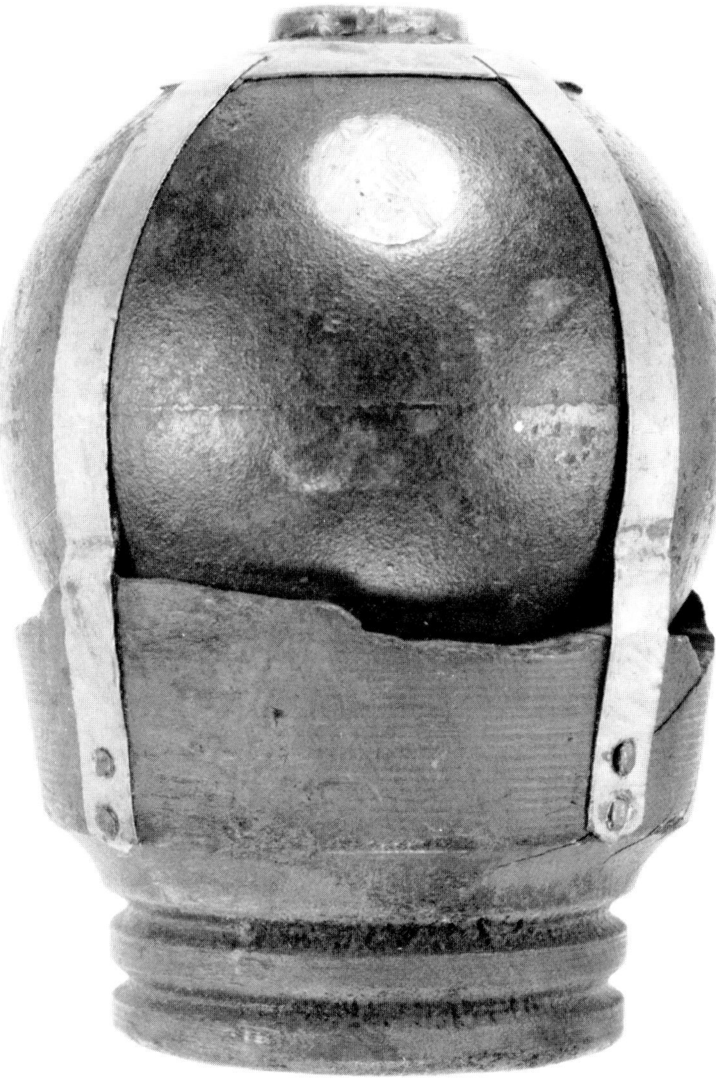

PATTERN: Spherical
DIAMETER: 4.52 inches including straps
GUN: 12-pounder smoothbore, 4.62-inch caliber
LENGTH: 6 1/8 inches including sabot
WEIGHT: 10 pounds, 9 ounces
TYPE: Case shot
SABOT: Wooden cup
FUSING: Copper fuse plug, paper time fuse

Comments: This lead side-loader ball is non-excavated and rests on its original wooden sabot. The twine used to tighten the four tin straps is missing. The sabot has two deep grooves to which the cartridge bag was tied. This sabot is longer than the specimen on page 51, figure III A-4. This projectile, when equipped with this sabot variety, was designed to be fired from the 12-pounder howitzer cannon. Note the crude mould seam around the horizontal center of the ball. The projectile contains iron case-shot balls that were inserted in the side hole. After insertion of the balls and matrix, a threaded lead bar was screwed into the side hole and sawed flush with the body.

Fig. III A-13

Confederate

PATTERN:	Spherical
DIAMETER:	4.49 inches
GUN:	12-pounder smoothbore, 4.62-inch caliber
LENGTH:	—
WEIGHT:	10 pounds, 8 ounces
TYPE:	Case shot
SABOT:	Wooden cup (missing)
FUSING:	Copper fuse plug, paper time fuse

Comments: Commonly called a lead side-loader ball, this type of projectile was commonly used in the Virginia campaigns. The lead side-loader plug sealed the hole through which the iron case-shot balls had been inserted. Iron balls were used due to the shortage of lead, which was needed for small arms ammunition. It is also found in the following calibers: 6-pounder (3.67-inch) and 24-pounder (5.82-inch) side-loaders. See page 17, figure II B-3, for a cross-section.

Fig. III A-14

Confederate

PATTERN:	Spherical
DIAMETER:	4.50 inches
GUN:	12-pounder smoothbore, 4.62-inch caliber
LENGTH:	—
WEIGHT:	9 pounds, 10 ounces
TYPE:	Case shot
SABOT:	Wooden sabot (missing)
FUSING:	Copper Bormann replacement fuse plug, paper time fuse

Comments: This smoothbore projectile has the Bormann replacement fuse plug and an iron side-loader plug. The Confederates discontinued the manufacturing of the Bormann fuse by 1863 and replaced it with an oversized copper fuse plug that screwed into the underplug threads. The iron plug measures .90" in diameter. The Bormann replacement fuse plug is also found in 6-pounder (3.67-inch caliber) and 24-pounder (5.82-inch caliber).

Fig. III A-15

Confederate

PATTERN: Spherical
DIAMETER: 4.50 inches
GUN: 12-pounder smoothbore, 4.62-inch caliber
LENGTH: —
WEIGHT: 9 pounds, 14 ounces
TYPE: Case shot
SABOT: Wooden cup (missing)
FUSING: Copper fuse plug, paper time fuse

Comments: The Confederates substituted iron case-shot balls for lead due to the shortage of lead needed for small arms ammunition. The Confederates could not drill into the iron balls to form the bursting cavity without great difficulty. The bursting cavity was formed by inserting a dowel into the fuse opening. The iron balls were inserted through the hole on the projectile's side, and the hot, liquefied matrix was poured in the side-loading hole. After the matrix cooled, the dowel was removed, and a lead, iron, copper, or brass plug was screwed into the loading hole. The powder was installed and the fuse adapter screwed in. Side-loader plugs made of copper or brass are uncommon. The copper side-loader plug measures .83" in diameter.

Fig. III A-16

Federal

PATTERN:	Spherical
DIAMETER:	5.65 inches
GUN:	24-pounder Coehorn mortar, 5.82-inch caliber
LENGTH:	—
WEIGHT:	16 pounds, 13 ounces
TYPE:	Shell
SABOT:	None
FUSING:	Wooden fuse plug, paper time fuse

Comments: Unlike the Confederate's Coehorn mortar ball, the Federals used four tin straps with loops on them to aid in loading and centering the projectile's fuse in the mortar. The Federal Coehorn mortar ball was manufactured without tong holes or projectile ears. The four straps cross underneath and wrap completely around the projectile. The pictured specimen is missing one of its straps. This style of Coehorn mortar projectile was used extensively in the Petersburg Campaign.

Fig. III A-17

Confederate

PATTERN: Spherical
DIAMETER: 5.63 inches
GUN: 24-pounder Coehorn mortar, 5.82-inch caliber
LENGTH: —
WEIGHT: 16 pounds, 4 ounces
TYPE: Shell
SABOT: None
FUSING: Wooden fuse plug, paper time fuse

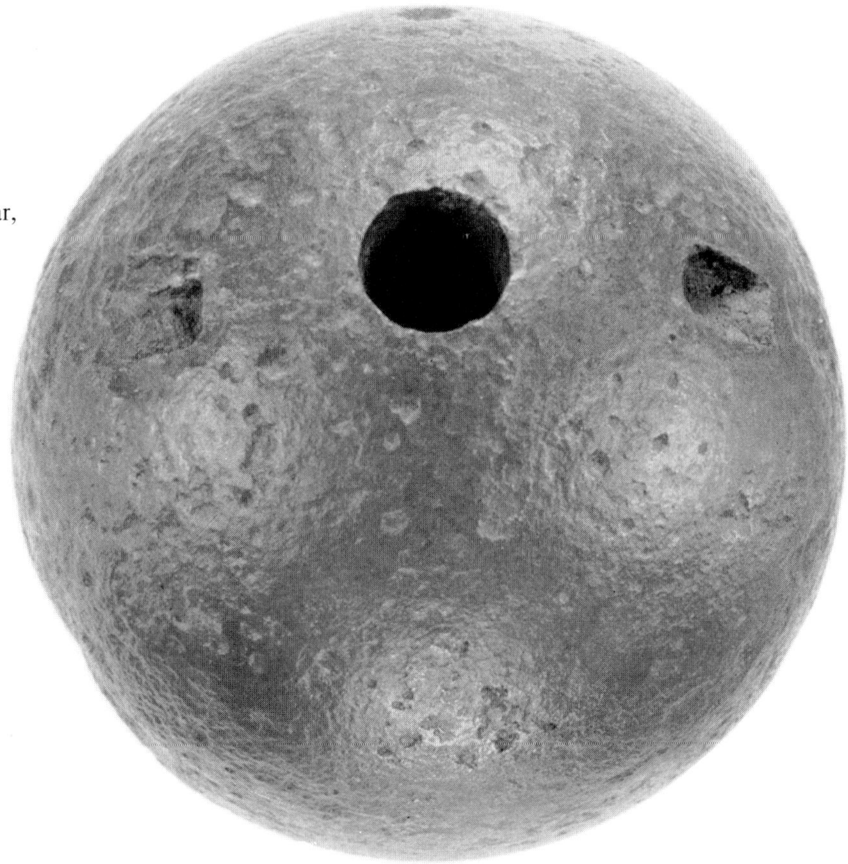

Comments: The Confederates added the tong holes on each side of the fuse to aid in loading and centering the projectile's fuse in the Coehorn mortar. Federal Coehorn mortar projectiles do not have tong holes or ears. See page 63, figure III A-16, for additional comments.

Fig. III A-18

Federal

PATTERN:	Conical
DIAMETER:	2.60 inches
GUN:	6-pounder Wiard rifle, 2.6-inch caliber
LENGTH:	6 5/16 inches
WEIGHT:	5 pounds, 9 ounces
TYPE:	Canister
SABOT:	Wooden cylinder
FUSING:	None

Comments: This non-excavated canister was intended for use in the 6-pounder, 2.6-inch Wiard rifle. It contains rows of iron balls packed with sawdust used as a filler. Specimens with lead balls have been recovered at Shiloh, Tennessee, and City Point, Virginia. The cartridge bag was tied onto the two deep grooves cut into the wooden sabot.

Fig. III A-19

Federal

PATTERN: Conical
DIAMETER: 2.55 inches
GUN: 6-pounder Wiard rifle, 2.6-inch caliber
LENGTH: 5 13/16 inches
WEIGHT: 5 pounds, 15 ounces
TYPE: Canister
SABOT: Lead cup
FUSING: None

Comments: This excavated canister was intended for use in the 6-pounder, 2.6-inch Wiard rifle. It has a thick lead base that is both recessed and concave. The sides are made of a thin lead sheeting. It contains rows of lead balls packed in sawdust, with a thin tin top crimped in place. Similar specimens are found in 3.67-inch and 3.8-inch caliber. On the base of the 3.67-inch lead canister, the word "BASE" appears in raised letters on the lead bottom. On the base of the 3.8-inch canister, the word "BASE 3.8" is in raised letters on the concave bottom. Federal government contracts show purchases of 2.6-inch canister from Hotchkiss & Sons.

Fig. III A-20

Federal

PATTERN: Conical
DIAMETER: 2.85 inches
GUN: 10-pounder Parrott rifle, 2.9-inch caliber
LENGTH: 7 inches
WEIGHT: 8 pounds
TYPE: Canister
SABOT: Wooden cylinder
FUSING: None

Comments: This non-excavated specimen is an example of the most common style of canister used by Federal artillery for the 10-pounder (2.9-inch caliber) cannon. The Confederates used a similar canister in this caliber. Most canisters have iron or lead balls stacked in rows inside, with sawdust used as a filler. The cartridge bag was tied onto the deep grooves cut into the wooden sabot. The canister walls were commonly made of tin to help prevent rusting while in service.

Fig. III A-21

Federal

PATTERN:	Hotchkiss
DIAMETER:	2.90 inches
GUN:	3-inch wrought iron (ordnance) rifle
LENGTH:	8 inches
WEIGHT:	7 pounds, 1 ounce
TYPE:	Canister
SABOT:	Lead cup
FUSING:	None

Comments: This non-excavated canister has "HOTCHKISS 3 IN JAN'Y 7 1862 PATENT" molded in raised letters on the bottom of the lead base. Specimens have been found with rows of lead or iron balls packed in sawdust. The sides are made of tinned sheet-iron, or what was ordinarily known as tin-plate. B.B. Hotchkiss patented this canister on January 7, 1862, patent #34,058.

Fig. III A-22

Federal

PATTERN:	Hotchkiss
DIAMETER:	3.57 inches
GUN:	20-pounder Parrott rifle, 3.67-inch caliber
LENGTH:	10 inches
WEIGHT:	14 pounds
TYPE:	Canister
SABOT:	Lead cup
FUSING:	None

Comments: This rare excavated canister contains rows of iron balls packed in sawdust. The tin sides protected the canister from rusting while in service. Molded in raised letters on the bottom of the lead base are the letters "HOTCHKISS 3 67/100 JAN. 7TH 1862 PATENT." Upon firing, the Hotchkiss canister would often take the grooves of the cannons rifling, creating an uneven dispersion of the canister balls.

Fig. III A-23

Federal

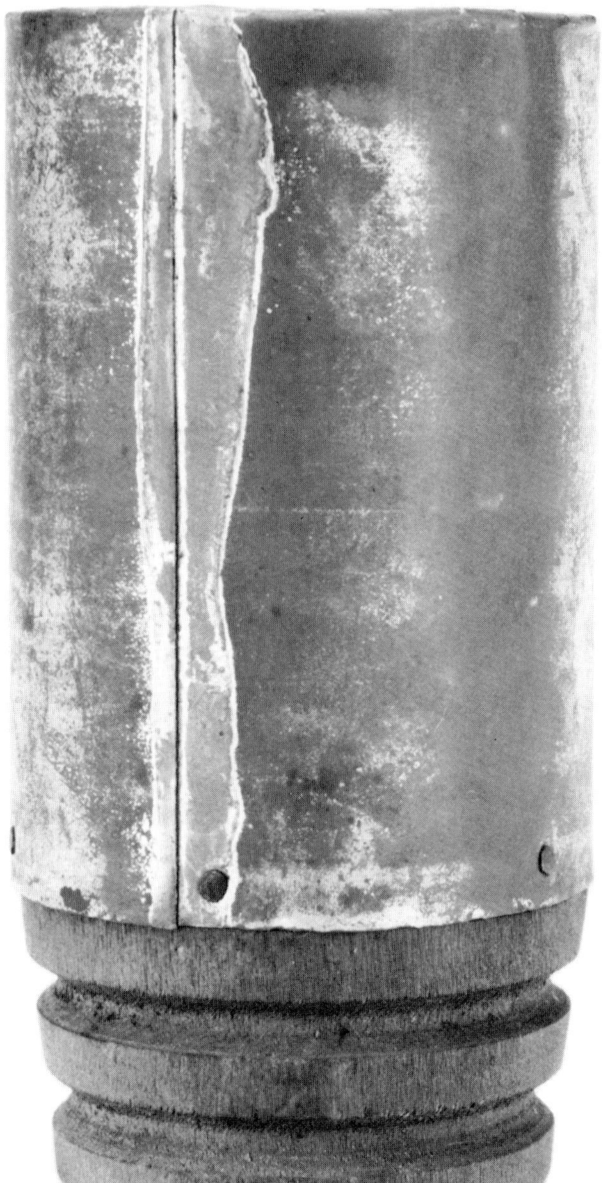

PATTERN: Spherical
DIAMETER: 3.59 inches
GUN: 6-pounder smoothbore, 3.67-inch caliber
LENGTH: 6 7/8 inches
WEIGHT: 7 pounds, 8 ounces
TYPE: Canister
SABOT: Wooden cylinder
FUSING: None

Comments: This non-excavated specimen is an example of the most common style of canister used in the 6-pounder smoothbore cannon. It contains rows of iron balls packed in sawdust. It is 3" shorter than the standard-style 20-pounder canister, also for a 3.67-inch bore. It could have also been used in the 20-pounder Parrott rifle, as it has the same size bore as the 6-pounder smoothbore. Both Northern and Southern forces used this size of canister. After attachment of the cartridge bag, which was tied to the deep grooves cut into the wooden sabot, the round was referred to as fixed ammunition.

Fig. III A-24

Federal

PATTERN: Sawyer
DIAMETER: 3.62 inches
GUN: Sawyer rifle, 3.67-inch caliber
LENGTH: 5 3/4 inches
WEIGHT: 3 pounds, 3 ounces (filler balls missing)
TYPE: Canister
SABOT: None
FUSING: None

Comments: This projectile is correctly referred to as Sawyer's solid-shot canister. A.M. Sawyer patented this ridged-wall canister on November 19, 1861, patent #33,754. It was designed not to expand into the guns rifling so that the expansion pattern of canister balls would be even. The Sawyer canister was designed so that it could be fired from a Sawyer rifle (3.67-inch caliber) or a 6-pounder smoothbore (3.67-inch caliber). This canister has three holes in the bottom (see base view), which allowed the propellant charge to help force out the canister balls. The thin metal top was held on with four pins. It contained rows of iron balls packed in sawdust. The majority of Sawyer canisters have been recovered at Port Hudson, Louisiana. It is also found in 3.4-inch caliber.

Fig. III A-25

Federal

PATTERN:	Sawyer
DIAMETER:	3.60 inches
GUN:	Sawyer rifle, 3.67-inch caliber
LENGTH:	5 13/16 inches
WEIGHT:	4 pounds, 4 ounces (filler balls missing)
TYPE:	Canister
SABOT:	None
FUSING:	None

Comments: This is correctly referred to as Sawyer's accelerating canister. The base view illustrates the unusual style of this canister. The hole in the base contained a powder train that ignited an explosive charge between the filler balls and the canister base. The charge ignited after leaving the gun, thus extending the canister's range. The ridged-wall Sawyer canister also acted as a projectile after the canister balls were dispersed. This was an added improvement, since the standard tin-sided canister shredded upon firing. The same comments apply as those on the previous Sawyer canister, page 71, figure III A-24..

Fig. III A-26

Federal

PATTERN: Conical
DIAMETER: 3.58 inches
GUN: 20-pounder Parrott rifle, 3.67-inch caliber
LENGTH: 9 7/8 inches
WEIGHT: 12 pounds, 7 ounces
TYPE: Canister
SABOT: Wooden cylinder
FUSING: None

Comments: This non-excavated specimen is the standard-style canister round for the 20-pounder Parrott rifle. It contains rows of iron balls packed in sawdust. It is 3" longer than the standard-style 6-pounder canister, also of 3.67-inch caliber. The cartridge bag was attached to the deep grooves of the wooden sabot. The authors have seen several 20-pounder canister specimens similar to the one pictured above.

Fig. III A-27

Federal

PATTERN:	James, Type I
DIAMETER:	3.63 inches
GUN:	6-pounder smoothbore or 3.67-inch caliber rifle
LENGTH:	8 inches
WEIGHT:	13 pounds, 3 ounces
TYPE:	Canister
SABOT:	Lead ring covered with a tin sleeve
FUSING:	None

Comments: This excavated specimen illustrates the construction of the James canister, showing the iron balls and tin covering. A James Type I sabot was used for the base. Refer to page 75, figure III A-28 (opposite), to see the James base only. The base has a tin-covered lead sabot over the iron ribs, and the sabot was covered with a lubricated canvas material. The tin sheeting over the lead helped keep the lead from fouling the grooves of the rifled cannon. The pictured specimen was found at Shiloh, Tennessee. The majority of James canisters have been unearthed in Western Theatre sites.

Fig. III A-28

Federal

PATTERN:	James, Type I
DIAMETER:	3.63 inches
GUN:	6-pounder smoothbore or 3.67-inch caliber rifle
LENGTH:	3 1/2 inches
WEIGHT:	4 pounds, 14 ounces
TYPE:	Canister
SABOT:	Lead ring covered with a tin sleeve (missing)
FUSING:	None

Comments: This is a view of a base from a James canister similar to that pictured on page 74, figure III A-27. It has been fired and, as usual, the lead covering has been stripped off. The canister sides were attached with six copper nails in the holes on the top of the base. The holes in the bottom sides of the base helped attach the lead band to the ribs. The base also acted as a projectile after leaving the cannon.

Fig. III A-29

Federal

PATTERN: Spherical
DIAMETER: 4.50 inches
GUN: 12-pounder smoothbore, 4.62-inch caliber
LENGTH: 8 inches
WEIGHT: 14 pounds, 11 ounces
TYPE: Canister
SABOT: Wooden cylinder
FUSING: None

Comments: This non-excavated specimen is the standard-style canister that both the Federal and Confederate forces used in the 12-pounder smoothbore cannon. Canister was fired at close range with a devastating effect on the enemy. See page 34, figure II B-20, for a reconstructed 12-pounder canister.

Fig. III A-30

Federal

PATTERN:	Spherical
DIAMETER:	4.42 inches
GUN:	12-pounder howitzer, 4.62-inch caliber
LENGTH:	8 3/4 inches
WEIGHT:	8 pounds, 8 ounces
TYPE:	Canister
SABOT:	Wooden cylinder
FUSING:	None

Comments: This non-excavated canister was designed for use in the 12-pounder smoothbore howitzer. The canister's wooden cylinder was tapered to fit the howitzer's tapered powder chamber. It contains rows of iron balls packed in sawdust. The illustrated disassembled view is of this actual canister. The single wide groove in the wooden base was for the attachment of the cartridge bag. The authors have examined several specimens of this type.

Fig. III A-31

Confederate

PATTERN:	Stand of grape
DIAMETER:	4.60 inches
GUN:	12-pounder smoothbore, 4.62-inch caliber
LENGTH:	5 13/16 inches to top plate
WEIGHT:	14 pounds, 6 ounces
TYPE:	Grapeshot
SABOT:	None
FUSING:	None

Comments: This non-excavated specimen is referred to as a stand of grape. When found separately in the field, the iron balls are often confused with canister rounds. The Federal government discontinued the 12-pounder stand of grape prior to 1861 and replaced it with canister. The shot found in canisters were large enough to be effective, and the canister possessed the advantage of striking a great many more points at one discharge than grape. Stands of grape are also found in the following field calibers: 18-pounder (5.3-inch) and 24-pounder (5.82-inch). Note the crude mold seams on the iron balls, which point to Confederate manufacture. Three complete stands were recovered in Vicksburg, Mississippi, and five from Savannah, Georgia. Complete excavated specimens are extremely rare.

Fig. III A-32

Pre-war Federal

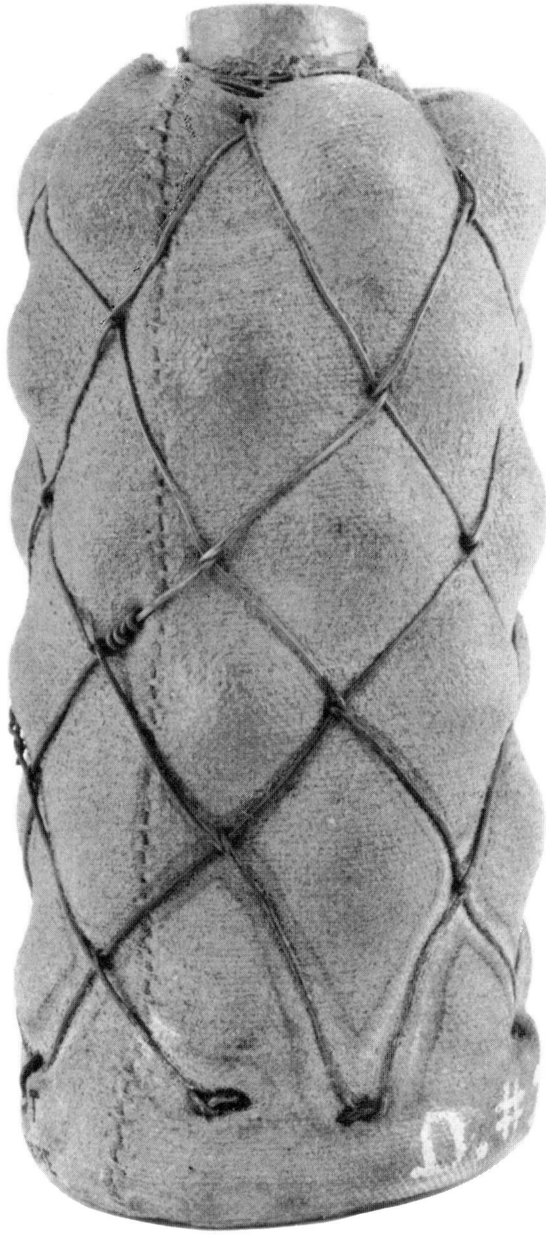

PATTERN:	Quilted grape
DIAMETER:	3.19 inches
GUN:	4-pounder smoothbore, 3.3-inch caliber
LENGTH:	6 1/2 inches
WEIGHT:	9 pounds
TYPE:	Grapeshot
SABOT:	Wooden disk
FUSING:	None

Comments: The authors feel that this non-excavated specimen is a pre-war quilted grape due to the 3.19-inch diameter for use in the 4-pounder (3.3-inch caliber) smoothbore cannon. It is too small for use in the 6-pounder (3.67-inch caliber) smoothbore cannon. Limited information is known about its use during the Civil War. Brass wire holds the canvas covering around the iron balls. The base and center post are made of wood.

Fig. III A-33

Confederate

PATTERN: Archer
DIAMETER: 2.94 inches
GUN: 3-inch rifle
LENGTH: 5 11/16 inches
WEIGHT: 7 pounds, 4 ounces
TYPE: Solid shot
SABOT: Lead band
FUSING: None

Comments: This projectile has two deep grooves on a tapered cone to help hold the lead sabot on during firing. It had a wooden cup on the base of the lead band to cushion the shock of the expansion, thereby helping to keep the sabot intact. Most fired Archers are found without the lead sabot. In the space between the lead band and the projectile body was a lubricated hemp rope. This lubricated rope was forced into the grooves of the cannon when the lead sabot expanded. The lubricant reduced friction and subsequent wear on the cannon. Similar specimens have been found in 3.3-inch caliber.

Fig. III A-34

Confederate

PATTERN:	Archer
DIAMETER:	2.94 inches
GUN:	3-inch rifle
LENGTH:	5 11/16 inches
WEIGHT:	5 pounds, 15 ounces
TYPE:	Solid shot
SABOT:	Lead band (missing)
FUSING:	None

Comments: This is the same pattern as the Archer on page 80, figure III A-33, except without the lead sabot. Note the two grooves on the tapered cone to hold the lead sabot on during expansion. Archer projectiles have often been recovered from the 1862 Shiloh, Tennessee, battlefield.

Fig. III A-35

Confederate

PATTERN: Archer
DIAMETER: 2.94 inches
GUN: 3-inch rifle
LENGTH: 5 11/16 inches
WEIGHT: 5 pounds, 12 ounces
TYPE: Solid shot
SABOT: Lead band (missing)
FUSING: None

Comments: This Archer is similar to the specimen pictured on page 81, figure III A-34, except that it has three deep grooves instead of two.

Fig. III A-36

Confederate

PATTERN: Archer
DIAMETER: 2.93 inches
GUN: 3-inch rifle
LENGTH: 6 inches
WEIGHT: 7 pounds, 13 ounces
TYPE: Solid shot
SABOT: Lead band
FUSING: None

Comments: This unfired projectile has two deep grooves to hold the lead sabot on the tapered base. Note the lubrication groove between the lead band and the body of the projectile. A lubricant was applied to the linen or canvas that covered the lead band. This helped prevent the bore of the cannon from being fouled with the lead from the sabot.

Fig. III A-37

Confederate

PATTERN:	Archer
DIAMETER:	2.94 inches
GUN:	3-inch rifle
LENGTH:	6 inches
WEIGHT:	7 pounds, 5 ounces
TYPE:	Solid shot
SABOT:	Lead band (missing)
FUSING:	None

Comments: This fired Archer, similar to that on page 83, figure III A-36, has two deep grooves to hold the lead sabot on. Note the casting flaw on the projectile's side.

Fig. III A-38

Confederate

PATTERN:	Archer
DIAMETER:	2.94 inches
GUN:	3-inch rifle
LENGTH:	6 inches
WEIGHT:	8 pounds, 7 ounces
TYPE:	Solid shot
SABOT:	Lead band
FUSING:	None

Comments: This unfired Archer variant has a small flat nose. It is less common than the pointed-nose variety. The tapered cone has two grooves that helped hold the lead band on during firing.

Fig. III A-39

Confederate

PATTERN: Archer
DIAMETER: 2.96 inches
GUN: 3-inch rifle
LENGTH: 6 1/2 inches
WEIGHT: 8 pounds, 10 ounces
TYPE: Solid shot
SABOT: Lead band
FUSING: None

Comments: This is a rare variant of the Archer patterns due to its rounded nose. It has four dimples on the tapered cone to hold the lead sabot on instead of the usual grooves. The lubrication groove is still visible on the fired sabot.

Fig. III A-40

Confederate

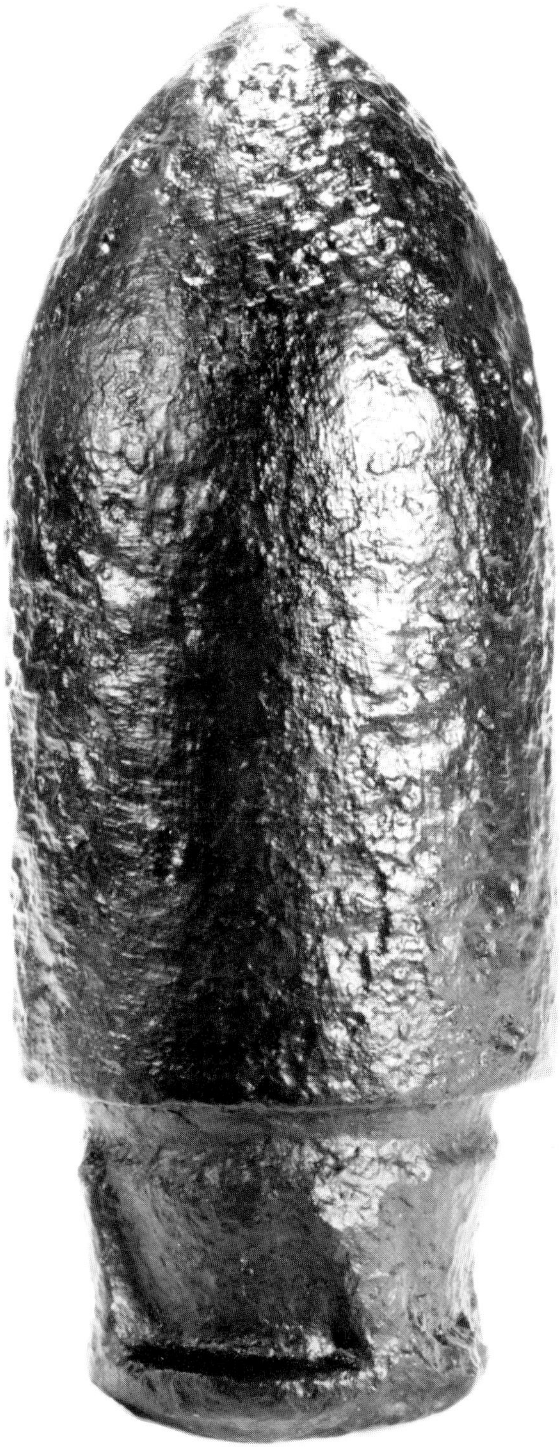

PATTERN: Archer
DIAMETER: 2.94 inches
GUN: 3-inch rifle
LENGTH: 7 1/4 inches
WEIGHT: 8 pounds, 15 ounces
TYPE: Solid shot
SABOT: Lead band (missing)
FUSING: None

Comments: This rare Trans-Mississippi Archer variant shows a tapered base with a groove and flange pattern, which aided in holding on the lead sabot. This pattern has been recovered in Louisiana.

Fig. III A-41

Confederate

PATTERN: Archer
DIAMETER: 2.94 inches
GUN: 3-inch rifle
LENGTH: 5 3/4 inches
WEIGHT: 6 pounds, 4 ounces
TYPE: Shell
SABOT: Lead band
FUSING: Wooden fuse plug, paper time fuse

Comments: This unfired Archer has two flame grooves on the iron projectile body. The lead band is shorter than those usually found on Archers. It has two grooves on the tapered cone to hold the sabot on. Note the wooden fuse plug in the nose of the projectile. Archer projectiles are named after Dr. Junius L. Archer of Bellona Arsenal in Virginia, where the majority of them were manufactured. However, this particular Archer was manufactured at Selma Arsenal.

Fig. III A-42

Confederate

PATTERN:	Archer
DIAMETER:	2.94 inches
GUN:	3-inch rifle
LENGTH:	5 1/2 inches
WEIGHT:	6 pounds, 8 ounces
TYPE:	Shell
SABOT:	Lead band
FUSING:	Wooden fuse plug, paper time fuse

Comments: This pattern is uncommon, since most Archer projectiles were of the solid-shot variety. Specimens have been noted with and without flame grooves. It has two deep grooves to help hold the lead sabot on the projectile.

Fig. III A-43

Confederate

PATTERN: Archer
DIAMETER: 2.97 inches
GUN: 3-inch rifle
LENGTH: 7 3/16 inches
WEIGHT: 8 pounds, 11 ounces
TYPE: Shell
SABOT: Lead band
FUSING: Wooden fuse plug, paper time fuse

Comments: This shell projectile is unfired and is of the long pattern in the Archer series. It has one groove and four vertical flanges to help hold the lead sabot on the projectile. An additional bottom ring or groove is thought to have been used to tie the cartridge bag onto the projectile body. Note the hemp rope that was greased to help lubricate the bore and cut down on the friction. This pattern variety has been recovered in Virginia.

Fig. III A-44

Confederate

PATTERN: Archer
DIAMETER: 2.94 inches
GUN: 3-inch rifle
LENGTH: 7 1/4 inches
WEIGHT: 7 pounds, 4 ounces
TYPE: Shell
SABOT: Lead band (missing)
FUSING: Wooden fuse plug, paper time fuse

Comments: This is a fired long-pattern Archer shell, the same pattern as page 90, figure III A-43, but without the lead sabot. Note the four vertical flanges and the tie-ring base.

Fig. III A-45

Confederate

PATTERN:	Archer
DIAMETER:	2.94 inches
GUN:	3-inch rifle
LENGTH:	7 1/8 inches
WEIGHT:	7 pounds, 10 ounces
TYPE:	Shell
SABOT:	Lead band (missing)
FUSING:	Confederate Bormann time fuse

Comments: The Bormann-fused Archer did not have enough burn time for the range of rifled artillery due to its 5 1/2-second time fuse. It has four vertically placed flanges and a groove to help hold on the lead sabot, which is missing on this specimen.

Fig. III A-46

Confederate

PATTERN: Archer
DIAMETER: 3.22 inches
GUN: 3.3-inch Confederate rifle
LENGTH: 6 1/2 inches
WEIGHT: 11 pounds, 15 ounces
TYPE: Solid shot
SABOT: Lead band
FUSING: None

Comments: The lower portion of the iron projectile body is cylindrical and not tapered as in the usual Archer pattern. Note that the lead sabot extends to the base of the projectile, which also is unusual for an Archer. The pictured specimen was recovered at Shiloh, Tennessee, and was fired by the 5th Company, Washington Artillery, of New Orleans.

Fig. III A-47

Confederate

PATTERN: Archer
DIAMETER: 3.22 inches
GUN: 3.3-inch Confederate rifle
LENGTH: 6 1/2 inches
WEIGHT: 8 pounds, 14 ounces
TYPE: Solid shot
SABOT: Lead band (missing)
FUSING: None

Comments: This fired Archer is the same variant as page 93, figure III A-46. There are three grooves on the straight cylindrical base.

Fig. III A-48

Confederate

PATTERN: Archer
DIAMETER: 3.26 inches
GUN: 3.3-inch Confederate rifle
LENGTH: 5 1/4 inches
WEIGHT: 7 pounds, 14 ounces
TYPE: Shell
SABOT: Lead band
FUSING: Wooden fuse plug, paper time fuse

Comments: This unfired Archer variety is found both with and without flame grooves. This is a seldom-seen size in the Archer series. There are three grooves and two small vertical flanges to help keep the sabot from slipping.

Fig. III A-49

Confederate

PATTERN:	Archer
DIAMETER:	3.23 inches
GUN:	3.3-inch Confederate rifle
LENGTH:	5 1/4 inches
WEIGHT:	5 pounds, 10 ounces
TYPE:	Shell
SABOT:	Lead band (missing)
FUSING:	Wooden fuse plug, paper time fuse

Comments: This fired projectile is of the same variety as the specimen on page 95, figure III A-48. It has two small vertical flame grooves placed on opposite sides of one another. There are two vertical flanges to help keep the sabot from turning on the base.

Fig. III A-50

Great Britain

PATTERN: Armstrong
DIAMETER: 3.12 inches
GUN: 12-pounder Armstrong breechloading rifle, 3-inch caliber
LENGTH: 8 9/16 inches
WEIGHT: 11 pounds, 1 ounce
TYPE: Shell
SABOT: Lead jacket
FUSING: Confederate copper fuse plug, paper time fuse

Comments: The Confederates had the exclusive use of this Armstrong-pattern projectile; it saw no service in the Federal forces. Indeed, it saw extremely limited use by the Southern forces. The Confederate fired one projectile of this pattern into General Abbot's Dutch Gap batteries in 1864. Abbot stated: "Had it not been for this circumstance, I should have supposed that none of the ordnance of this gun-maker was ever used by either army in Virginia." A small number of unfired Armstrong projectiles were found in a Confederate fort at High Bridge, Virginia.

This projectile is commonly known as the wasp-waist Armstrong. The lead sabot is secured to the iron projectile body by five raised rings around the projectile. The thick lead band absorbed most of the bursting charge explosion, thus making any fragmentation very incomplete.

Fig. III A-51

Great Britain

PATTERN:	Britten
DIAMETER:	2.45 inches
GUN:	2.5-inch Blakely rifle
LENGTH:	5 1/8 inches
WEIGHT:	6 pounds, 3 ounces
TYPE:	Solid shot
SABOT:	Lead cup
FUSING:	None

Comments: This 2.5-inch Britten was used exclusively by the Confederates for their imported British Blakely rifled cannon. Note the six lands and grooves of the Blakely sawtooth rifling on the lead sabot. The pictured projectile was recovered at Manassas, Virginia.

Fig. III A-52

Great Britain

PATTERN:	Britten
DIAMETER:	2.87 inches
GUN:	3-inch wrought iron (ordnance) rifle
LENGTH:	5 1/8 inches
WEIGHT:	7 pounds, 2 ounces
TYPE:	Solid shot
SABOT:	Lead cup
FUSING:	None

Comments: This projectile is a rare imported British Britten solid shot fired by the Federal artillery in Helena, Arkansas, the only known site for battlefield recoveries. Note that it has seven lands and grooves on the lead sabot, indicating that it was fired from a 3-inch wrought iron (ordnance) rifle. The sabot is swelled where it expanded into the rifling of the cannon. This projectile pattern was intended for use in a 2.9-inch rifle. This is an example of imported ammunition used by the Federal forces.

Fig. III A-53

Great Britain

PATTERN:	Britten
DIAMETER:	2.86 inches
GUN:	3-inch wrought iron (ordnance) rifle
LENGTH:	5 inches
WEIGHT:	6 pounds, 2 ounces
TYPE:	Segmented shell
SABOT:	Lead cup
FUSING:	Threaded fuse missing

Comments: This is the shell version of the solid shot on page 99, figure III A-52. Bashley Britten patented this projectile's segmented interior construction dated March 8, 1861, British patent #585. His patent describes the method of manufacturing the interior: "I construct this projectile in the following manner:—I make a core of loam or sand of such form as to fill the interior space. Round this core I place numerous pieces shaped and arranged somewhat like the natural divisions of an orange. Each of these pieces is cast with several cores or plates partially dividing them so as to weaken them at those places, that they may easily break up when the shell explodes. The surfaces of these pieces are dressed over with a coating of loam or such like material, and they are bound round with a wire, or otherwise secured round the inner core of sand .... Round this combined core I cast the external shell, which may be of iron or other suitable metal."

Fig. III A-54

Great Britain

PATTERN: Britten
DIAMETER: 3.45 inches
GUN: 12-pounder Blakely rifle, 3.5-inch caliber
LENGTH: 5 7/8 inches
WEIGHT: 11 pounds, 8 ounces
TYPE: Solid shot
SABOT: Lead cup
FUSING: None

Comments: This is the mate to page 99, figure III A-52. It is the standard-style shot for the 12-pounder Blakely rifled cannon. Fired specimens have almost always been found with their lead sabots intact. According to Britten's British patent #1,740, dated August 1, 1855, "The method of coating the projectiles with soft metal in such a manner as to stand the explosive force of the powder is as follows:—The iron is first coated with zinc by the process commonly known as the galvanising process, and while sufficiently hot to keep the zinc in a fused state on its surface, it is plunged into a mould or vessel of suitable form containing the lead or other soft metal in a fused state, and then allowed to get cold."

Fig. III A-55

Great Britain

PATTERN:	Britten
DIAMETER:	3.44 inches
GUN:	12-pounder Blakely rifle, 3.5-inch caliber
LENGTH:	7 1/4 inches
WEIGHT:	12 pounds, 2 ounces
TYPE:	Segmented shell
SABOT:	Lead cup
FUSING:	Britten percussion

Comments: This is the shell version of the projectile on page 100, figure III A-53, and it is the most common Britten shell manufactured. It was patented by Bashley Britten, British patent #585 dated March 8, 1861, for segmented interior construction. The interior has nine long wedges of iron with lines of weakness cast in each wedge. Ideally, the bursting charge broke the long wedges into 54 small fragments, and the outer shell body into fragments as well. Note the Britten percussion fuse with its distinctive brass anvil cap. See the comments on page 100 for a description of the manufacture of the interior.

Fig. III A-56

Confederate

PATTERN:	Brooke
DIAMETER:	2.94 inches
GUN:	3-inch Parrott rifle
LENGTH:	7 7/8 inches
WEIGHT:	8 pounds, 12 ounces
TYPE:	Shell
SABOT:	Copper rachet plate
FUSING:	Wooden fuse plug, paper time fuse

Comments: This pattern is named after its inventor, Commander John M. Brooke, C.S.N. It is commonly called a rebated Brooke. There are six ratchets in the copper sabot that correspond to the six cast-in ratchets on the shell base. A single bolt holds the sabot on. The purpose of the rebated section was to reduce the amount of surface area that needed to be turned on a lathe. Specimens have been recovered in Nashville, Tennessee, and in the Atlanta, Georgia, area.

Fig. III A-57

Confederate

PATTERN: Brooke
DIAMETER: 2.85 inches
GUN: 10-pounder Parrott rifle, 2.9-inch caliber
LENGTH: 9 inches
WEIGHT: 10 pounds, 1 ounce
TYPE: Shell
SABOT: Copper rachet plate
FUSING: Wooden fuse plug, paper time fuse

Comments: This is the most common 10-pounder Brooke projectile. It was turned on a lathe from the sabot to the ogive. Lathing was needed to turn the casting down to the dimensions needed to properly fit the bore of the cannon. Stamped on the copper sabot is the number "8." The sabots have also been stamped on the side with a "G" and a "D." It is also found in a shorter pattern.

Fig. III A-58

Confederate

PATTERN:	Brooke
DIAMETER:	2.84 inches
GUN:	10-pounder Parrott rifle, 2.9-inch caliber
LENGTH:	9 1/4 inches
WEIGHT:	10 pounds, 8 ounces
TYPE:	Shell
SABOT:	Copper rachet plate
FUSING:	Copper fuse plug, paper time fuse

Comments: This sabot is pre-cast with flanges to fit the gun's grooves. The six ratchets on the copper plate correspond to the inclined ratchets on the projectile base. When fired, the ratchets tightened against each other, equally absorbing the energy and thus increasing the sabot stability. The copper time fuse plug is slotted. This projectiles was used exclusively by Forrest's Artillery and only in north Alabama and Mississippi.

Fig. III A-59

Confederate

PATTERN:	Brooke
DIAMETER:	3.63 inches
GUN:	20-pounder Parrott rifle, 3.67-inch caliber
LENGTH:	10 3/16 inches
WEIGHT:	16 pounds, 7 ounces
TYPE:	Shell
SABOT:	Copper ratchet plate
FUSING:	Copper fuse plug, paper time fuse

Comments: This 20-pounder Brooke projectile variety is seldom seen. It has the letter "D" stamped on the copper sabot. Brooke projectiles are turned on a lathe to the bore dimensions required. Note that the lathe marks are clearly visible on the sabot. There are five ratchets on the sabot, unlike the 10-pounder series. Also, the letter "G" is found stamped on the sabot. A good number of these projectiles have been recovered from Atlanta Campaign sites.

Fig. III A-60

Confederate

PATTERN: Broun
DIAMETER: 2.94 inches
GUN: 3-inch rifle
LENGTH: 7 1/2 inches
WEIGHT: 9 pounds, 4 ounces
TYPE: Shell
SABOT: Copper ring
FUSING: Wooden fuse plug, paper time fuse

Comments: This pattern is found mainly on late-war battlefields. The bottom bourrelet was omitted on Broun projectiles. Only the top bourrelet and the sabot needed to be machined to tolerance. This is the short pattern; the longer version was approximately 3/4 of an inch longer. There are two saw cuts to aid in the expansion of the sabot into the grooves of the gun. Most fired specimens have the thin sabot blown off, revealing the slightly rounded base. The letter "T" has been found stamped on the sabot on other specimens.

Fig. III A-61

Confederate

PATTERN: Broun
DIAMETER: 2.94 inches
GUN: 3-inch rifle
LENGTH: 8 3/16 inches
WEIGHT: 9 pounds, 2 ounces
TYPE: Shell (possibly segmented)
SABOT: Copper ring
FUSING: Confederate copper fuse plug, paper time fuse

Comments: Broun projectiles are named after their inventor, Lieutenant Colonel William Leroy Broun (commander of the Richmond Arsenal beginning in June 1863), who modified the then-popular Read projectiles. Examples such as this with the copper time fuse plug are uncommon. Most Broun projectiles recovered have a wooden fuse plug for a paper time fuse. The copper sabot is .90" high and has two saw cuts to help aid in the expansion in the rifling of the cannon. The sabots were nearly always blown off during firing. The sabot ring was cast into notches in the exterior of the base, which proved to be a less than effective method of attachment.

Fig. III A-62

Confederate

PATTERN:	Broun
DIAMETER:	2.94 inches
GUN:	3-inch rifle
LENGTH:	8 1/4 inches
WEIGHT:	9 pounds, 1 ounce
TYPE:	Segmented shell
SABOT:	Copper ring
FUSING:	Wooden fuse plug, paper time fuse

Comments: The interior is in the form of a six-sided prism. The stars have horizontal lines of weakness designed to break the shell into 24 fragments, plus the nose and base. This projectile was manufactured at Selma Arsenal in Alabama. Note that the projectile has only one raised bearing ring, which is typical of Broun projectiles.

Fig. III A-63

Confederate

PATTERN:	Broun
DIAMETER:	3.61 inches
GUN:	3.67-inch caliber rifle
LENGTH:	5 1/2 inches
WEIGHT:	8 pounds, 12 ounces
TYPE:	Segmented shell
SABOT:	Copper ring
FUSING:	Wooden fuse plug, paper time fuse

Comments: It is believed that this short-variety projectile tumbled in flight due to its length-to-caliber ratio. This projectile has twelve rectangular sections with lines of weakness cast in. The interior was designed to fragment into equal pieces, with the exception of the nose and base. The base is slightly rounded. Note the lathe marks on the bourrelet and sabot.

Fig. III A-64

Confederate

PATTERN:	Carkeet
DIAMETER:	1.94 inches
GUN:	2-pounder Carkeet smoothbore, 2-inch caliber
LENGTH:	3 13/16 inches
WEIGHT:	2 pounds, 2 ounces
TYPE:	Solid shot
SABOT:	None
FUSING:	None

Comments: James Hill Carkeet was issued a Confederate patent for a 2-pounder Carkeet gun. To the authors' knowledge, the projectile pictured was not patented. This projectile used two air vanes as a system of rotation. The rotary motion of the projectile occurred when the air passed over the two vanes, producing a left-hand twist. The concave base helped keep the projectile's weight evenly distributed. Several specimens were recovered near Selma, Alabama.

Fig. III A-65

Federal

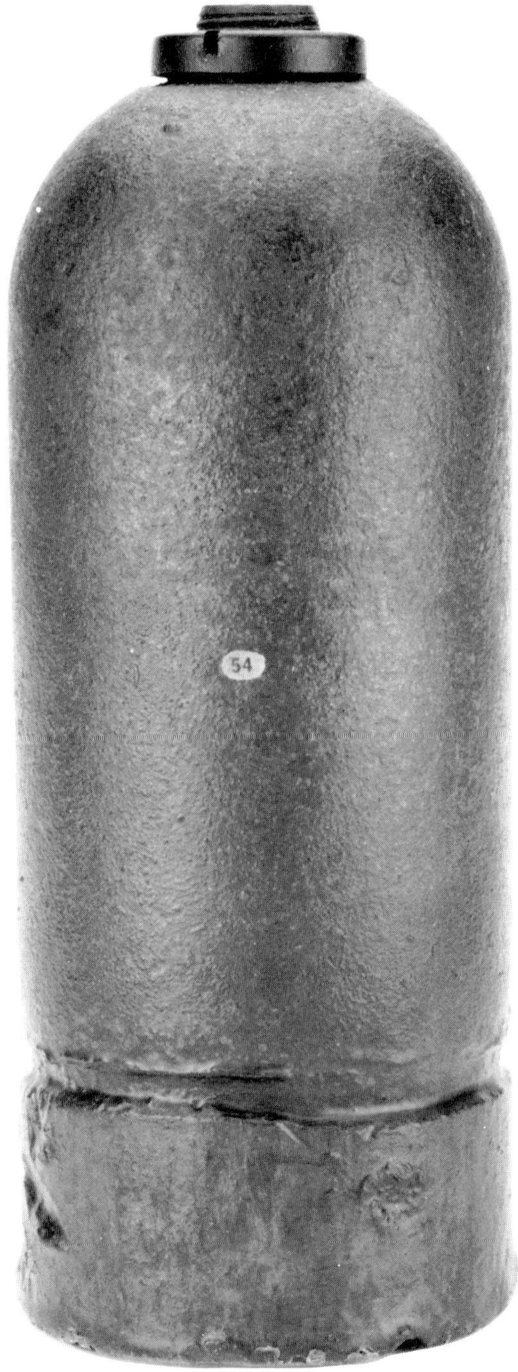

PATTERN: Dyer
DIAMETER: 2.91 inches
GUN: 3-inch wrought iron (ordnance) rifle
LENGTH: 7 3/8 inches
WEIGHT: 8 pounds, 5 ounces
TYPE: Shell
SABOT: Lead cup
FUSING: Federal Schenkl percussion

Comments: This has been mistakenly called a Confederate Burton projectile. Federal government records indicate that William Sellers & Company manufactured 107,230 3-inch Dyer projectiles under their Federal ordnance contracts from September 26, 1861, through May 24, 1862. The pattern is found in case shot and common shell. The lead sabot is found both with and without flame grooves. The majority of the specimens were fired from Federal artillery positions, although some were captured. The Schenkl fuse has the patent date stamped on the face.

Fig. III A-67

Federal

PATTERN:	Dyer
DIAMETER:	2.93 inches
GUN:	3-inch wrought iron (ordnance) rifle
LENGTH:	7 1/8 inches
WEIGHT:	10 pounds, 9 ounces
TYPE:	Case shot
SABOT:	Lead cup
FUSING:	Zinc fuse plug, paper time fuse

Comments: This pattern has four flame grooves and a lubrication groove above the lead sabot. Often the lead sabot of the 3-inch Dyer projectile became separated during firing. According to a letter found in the National Archives written by Major J.G. Benton to General A.B. Dyer and dated October 28, 1864, this problem was corrected: "... the old cups can be easily removed and good ones put on at a cost to the Government of 15 cents. There are about 55,000 of these projectiles at this Arsenal and I think that they can be made serviceable by Mr. Taylor's plan ... Forty projectiles arranged in this way were fired without a single failure or detachment of the sabot." See page 113, figure III A-66, for details on one method used to repair the lead sabots.

Fig. III A-66

Federal

PATTERN:	Dyer
DIAMETER:	2.94 inches
GUN:	3-inch wrought iron (ordnance) rifle
LENGTH:	7 1/8 inches
WEIGHT:	10 pounds, 8 ounces
TYPE:	Case shot
SABOT:	Lead cup
FUSING:	Zinc fuse plug, paper time fuse

Comments: This specimen does not have flame grooves. As can be seen in the base view above, a pair of crossed tin straps were soldered to the iron base of the projectile through a hole in the lead base. This was an attempt to help correct the problem of the sabots falling off the projectile after leaving the cannon. This specimen was fired from a Federal 3-inch wrought iron (ordnance) rifle. See page 20, figure II B-6, for a cross-sectional view of a shell. For more information about the Dyer projectiles, we refer you to *North South Trader's CIVIL WAR* magazine, Vol. XX No. 6 (Nov.-Dec. 1993).

Fig. III A-68

Federal converted to Confederate

PATTERN:	Dyer
DIAMETER:	2.93 inches
GUN:	3-inch wrought iron (ordnance) rifle
LENGTH:	7 1/4 inches
WEIGHT:	10 pounds, 1 ounce
TYPE:	Case shot
SABOT:	Lead cup
FUSING:	Confederate copper fuse plug, paper time fuse

Comments: This unfired specimen was captured by the Confederates, reworked, fitted with a Confederate copper fuse plug, and drilled on the side to accept iron case-shot balls. Only a few side-loader specimens have been noted. The side-loader plug is made of lead. There are two tin crossed straps on the projectile base and on the lead sabot (for a base view, see page 113, figure III A-66). The Federal zinc fuse plug and the Confederate copper fuse plug have the same thread count and diameter. The fuses would be interchangeable.

Fig. III A-69

Federal converted to Confederate

PATTERN: Dyer
DIAMETER: 2.92 inches
GUN: 3-inch rifle
LENGTH: 7 1/4 inches
WEIGHT: 10 pounds, 2 ounces
TYPE: Case shot
SABOT: Lead cup
FUSING: Confederate copper fuse plug, paper time fuse

Comments: The lead side-loader plug is 7/8" in diameter. The lead sabot has four flame grooves. There are no crossed straps on the base of the sabot on this specimen. The same comments apply as those on page 115, figure III A-68.

Fig. III A-70

Federal

PATTERN:	Dyer
DIAMETER:	2.94 inches
GUN:	3-inch wrought iron (ordnance) rifle
LENGTH:	7 1/2 inches
WEIGHT:	10 pounds
TYPE:	Case shot
SABOT:	Lead cup
FUSING:	Zinc fuse plug, paper time fuse

Comments: This specimen is more pointed than the previous Dyer pattern and is therefore less common. It is found in both shell and case shot. There are four flame grooves and a lubrication groove above the lead sabot. The sabot is similar to that on page 114, figure III A-67.

Fig. III A-71

Confederate

PATTERN:	Ellsworth
DIAMETER:	1.47 inches
GUN:	1.5-inch Ellsworth breechloading rifle
LENGTH:	3 1/4 inch
WEIGHT:	1 pound, 2 ounces
TYPE:	Solid shot
SABOT:	Tallow-soaked cord
FUSING:	None

Comments: This specimen was reportedly found in Alabama at the site of the Selma Arsenal, where it was manufactured for the captured Federal Ellsworth cannon. The projectile base is slightly tapered for the probable use of a wound tallow-soaked cord that acted as a sabot.

Fig. III A-72

Federal

PATTERN: Ellsworth
DIAMETER: 1.45 inches
GUN: 1.5-inch Ellsworth breechloading rifle
LENGTH: 3 3/16 inches
WEIGHT: 1 pound, 3 ounces
TYPE: Solid shot
SABOT: Flexible plate (missing)
FUSING: None

Comments: There have been several battlefield specimens of this pattern recovered. Due to the projectile's weight and size, it is believed to be for the Ellsworth breechloading rifle. There is a square hole in the base approximately 1/4" wide and deep. One specimen was found at Fort Hindman, Kentucky.

Fig. III A-73

Ellsworth (Tentative)

PATTERN:	Ellsworth
DIAMETER:	1.49 inches
GUN:	Probably 1.5-inch Ellsworth breechloading rifle
LENGTH:	5 1/4 inches
WEIGHT:	2 pounds, 4 ounces
TYPE:	Solid shot
SABOT:	Lead band
FUSING:	None

Comments: This specimen is possibly a projectile for the Ellsworth breechloading rifle due to its caliber and weight. There is a space between the lead band and the cylindrical portion of the projectile body. This allowed the propellant charge to force the lead band into the rifling grooves. The specimen pictured was found in West Virginia.

Fig. III A-74

Federal

PATTERN: Hotchkiss
DIAMETER: 2.54 inches
GUN: 6-pounder Wiard rifle, 2.6-inch caliber
LENGTH: 6 inches
WEIGHT: 6 pounds, 3 ounces
TYPE: Solid shot
SABOT: Lead band
FUSING: None

Comments: This was the smallest caliber Hotchkiss pattern manufactured, and it is the rarest variety of the 2.6-inch Wiard rifle Hotchkiss projectiles. Hotchkiss projectiles were manufactured in three pieces: a nose, a lead band, and a base cup. See page 126, figure III A-79, for a disassembled view of a similar projectile.

Fig. III A-75

Federal

PATTERN: Hotchkiss
DIAMETER: 2.53 inches
GUN: 6-pounder Wiard rifle, 2.6-inch caliber
LENGTH: 5 7/8 inches
WEIGHT: 5 pounds, 4 ounces
TYPE: Shell
SABOT: Lead band
FUSING: Wooden fuse plug, paper time fuse

Comments: This non-excavated specimen still retains some of the original arsenal paint. On the base is cast "HOTCHKISS PAT. OCT 9, 1855," which signifies the date on which the patent was applied for. It was subsequently issued on October 16, 1855, patent #13,679. The Hotchkiss patent date is on the base. It has three flame grooves that enabled the propellant charge flame to more readily travel past the body and light the paper time fuse. Hotchkiss projectiles of 2.6-inch caliber are scarce.

Fig. III A-76

Federal

PATTERN:	Hotchkiss
DIAMETER:	2.53 inches
GUN:	6-pounder Wiard rifle, 2.6-inch caliber
LENGTH:	6 9/16 inches
WEIGHT:	5 pounds, 13 ounces
TYPE:	Shell
SABOT:	Lead band
FUSING:	Percussion, iron anvil cap

Comments: This projectile has the Hotchkiss patent date cast into the bottom of the base cup. Smith and Stetson patented this style of fuse on March 25, 1862, patent #34,788. The fuse pictured is a type developed at West Point, New York, and is often referred to as a West Point percussion type.

Fig. III A-77

Confederate

PATTERN: Hotchkiss
DIAMETER: 2.80 inches
GUN: Possibly 2.9-inch rifle
LENGTH: 5 15/16 inches (nose only)
WEIGHT: 6 pounds
(base cup and sabot missing)
TYPE: Solid shot
SABOT: Lead band (missing)
FUSING: None

Comments: The specimen pictured is the only one known to the authors, who believe it to be a Confederate copy of the Hotchkiss pattern. The nose has a dimple in the top and a casting flaw in the base. It appears to be too small in diameter for the 2.9-inch rifle, and it is too large for the 2.75-inch rifle. This specimen was recovered from Shiloh, Tennessee.

Fig. III A-78

Federal

PATTERN: Hotchkiss
DIAMETER: 2.94 inches
GUN: 3-inch wrought iron (ordnance) rifle
LENGTH: 7 1/4 inches
WEIGHT: 10 pounds, 1 ounce
TYPE: Solid shot
SABOT: Lead band
FUSING: None

Comments: Andrew Hotchkiss patented this pattern of projectile on October 16, 1855, patent #13,679. Andrew died in 1858 and his brother Benjamin patented an improved pattern on May 14, 1861, patent #32,293. See page 126, figure III A-79, for a disassembled view showing the construction of the Hotchkiss projectile.

Fig. III A-79

Federal

PATTERN: Hotchkiss
DIAMETER: 2.94 inches
GUN: 3-inch wrought iron (ordnancc) rifle
LENGTH: —
WEIGHT: —
TYPE: Solid shot
SABOT: Lead band
FUSING: None

Comments: This disassembled view shows the three-piece construction of the Hotchkiss projectile (the nose, the lead band, and the base cup). Upon the ignition of the gun, the base cup was driven along the cylindrical base of the projectile body, expanding the lead sabot into the rifling. Due to the expansion of the lead band, the 3-inch wrought iron (ordnance) rifle's seven lands and grooves allowed a more even distribution of the stress. The 3-inch Parrott rifle had only three lands and grooves, which was not enough to distribute the stress correctly.

Fig. III A-80

Federal

PATTERN: Hotchkiss
DIAMETER: 2.94 inches
GUN: 3-inch wrought iron (ordnance) rifle
LENGTH: 7 1/8 inches
WEIGHT: 9 pounds
TYPE: Case shot
SABOT: Lead band
FUSING: Lead fuse plug, paper time fuse

Comments: This unfired Hotchkiss projectile is an early variety with no flame grooves. When expanded, the lead band sometimes sealed the bore so well that no flame could pass, thereby keeping the shell's paper time fuse from igniting. This led to the development of flame grooves, patented by Benjamin Hotchkiss on June 7, 1864, patent #43,027. Earlier Hotchkiss shells have no flame grooves and are found with wooden or metal fuse adapters. There are two spanner holes on the top of the fuse plug for installation, rather than the usual slots.

Fig. III A-81

Federal

PATTERN: Hotchkiss
DIAMETER: 2.94 inches
GUN: 3-inch wrought iron (ordnance) rifle
LENGTH: 6 11/16 inches
WEIGHT: 8 pounds, 10 ounces
TYPE: Shell
SABOT: Lead band
FUSING: Wooden fuse plug, paper time fuse

Comments: This is an early Hotchkiss projectile with three flame grooves that are smaller than those usually found. A wooden fuse plug was also uncommon for this projectile in 3-inch caliber. The Hotchkiss patent date was cast into the bottom of the base cup. See page 136, figure III A-89, for an example of the patent date. This Hotchkiss variety is found mostly on Virginia battlefields.

Fig. III A-82

Federal

PATTERN: Hotchkiss
DIAMETER: 2.94 inches
GUN: 3-inch wrought iron (ordnance) rifle
LENGTH: 6 7/8 inches
WEIGHT: 8 pounds, 3 ounces
TYPE: Shell
SABOT: Lead band
FUSING: Hotchkiss brass fuse plug, paper time fuse

Comments: More of this particular type of 3-inch Hotchkiss shell was manufactured than any other style of Hotchkiss during the Civil War. Specimens can be found with the Hotchkiss patent on the base cup. Note the original black arsenal paint on this unfired non-excavated specimen.

Fig. III A-83

Federal

PATTERN: Hotchkiss
DIAMETER: 2.94 inches
GUN: 3-inch wrought iron (ordnance) rifle
LENGTH: 6 7/8 inches
WEIGHT: 9 pounds, 2 ounces
TYPE: Case shot
SABOT: Lead band
FUSING: Hotchkiss brass fuse plug, paper time fuse

Comments: The case-shot variety has a more rounded nose than the shell version in this pattern. The Hotchkiss patent date is cast into the base cup. Note the flame grooves needed to help communicate the propellant flame to the paper time fuse. This specimen was recovered from the battlefield of Kennesaw Mountain, Georgia. See page 22, figure II B-8, for a cross-sectional view of the interior construction.

Fig. III A-84

Federal

PATTERN:	Hotchkiss
DIAMETER:	2.94 inches
GUN:	3-inch wrought iron (ordnance) rifle
LENGTH:	6 5/8 inches
WEIGHT:	8 pounds, 7 ounces
TYPE:	Shell
SABOT:	Lead band
FUSING:	Hotchkiss percussion, brass

Comments: No flame grooves are needed for percussion-fused shells. These 3-inch Hotchkiss projectiles were designed for use in the 3-inch wrought iron (ordnance) rifle and not the 3-inch Parrott rifle. B.B. Hotchkiss patented the brass percussion fuse on February 24, 1863, patent #37,756.

Fig. III A-85

Federal

PATTERN: Hotchkiss
DIAMETER: 2.94 inches
GUN: 3-inch wrought iron (ordnance) rifle
LENGTH: 6 5/16 inches
WEIGHT: 9 pounds, 8 ounces
TYPE: Shell
SABOT: Lead band
FUSING: Hotchkiss brass fuse plug, paper time fuse

Comments: This projectile is unfired and is of a late-war production pattern. Its flat nose is fitted with a standard brass Hotchkiss time fuse adapter. It has three flame grooves and a slightly rounded base. This projectile was replaced by the 14-second Hotchkiss-Bormann time fuse version. Specimens have been recovered in Nashville, Tennessee, and at Spanish Fort, Alabama.

Fig. III A-86

Federal

PATTERN: Hotchkiss
DIAMETER: 2.94 inches
GUN: 3-inch wrought iron (ordnance) rifle
LENGTH: 6 1/4 inches
WEIGHT: 9 pounds, 1 ounce
TYPE: Case shot
SABOT: Lead band
FUSING: Hotchkiss combination percussion and 14-second Bormann time fuse

Comments: Benjamin Hotchkiss patented this improvement for the Hotchkiss projectile on May 16, 1865, patent #47,725. A wooden disk was inserted between the base cup and the base of the nose to decrease the risk of premature fracture of the shell during initial firing. The insufficient burning time of the standard-style 5 1/4-second Bormann time fuse led to the development of the 14-second version for rifled ordnance. The protective cover of the Hotchkiss combination fuse pictured here was not removed prior to firing. See page 23, figure II B-9, for a cross-sectional view. Note the seven lands and grooves of the 3-inch wrought iron (ordnance) rifle on the lead sabot.

Fig. III A-87

Federal

PATTERN: Hotchkiss
DIAMETER: 3.42 inches
GUN: 12-pounder Blakely rifle, 3.5-inch caliber
LENGTH: 6 3/8 inches
WEIGHT: 12 pounds, 3 ounces
TYPE: Case shot
SABOT: Lead band
FUSING: Hotchkiss brass fuse plug, paper time fuse

Comments: The Hotchkiss pattern pictured above is uncommon in this caliber. Notice the nice rifling on the sabot, consisting of the seven lands and grooves of the 3.5-inch Blakely rifle.

Fig. III A-88

Federal

PATTERN:	Hotchkiss
DIAMETER:	3.59 inches
GUN:	3.67-inch caliber rifle
LENGTH:	6 11/16 inches
WEIGHT:	12 pounds, 14 ounces
TYPE:	Solid shot
SABOT:	Lead band
FUSING:	None

Comments: This is the standard solid shot for the 3.67-inch caliber rifle. This variety of projectile nearly always kept its sabot after firing. A large number have been recovered from Vicksburg and Port Gibson, Mississippi.

Fig. III A-89

Federal

PATTERN: Hotchkiss
DIAMETER: 3.58 inches
GUN: 3.67-inch caliber rifle
LENGTH: 7 inches
WEIGHT: 12 pounds
TYPE: Shell
SABOT: Lead band
FUSING: Percussion, iron anvil cap

Comments: On March 25, 1862, C.W. Smith and T.D. Stetson patented the fuse that is in this pictured specimen, patent #34,788. The lead sabot still has the impression of the linen fabric that was used to keep the lead from fouling the bore of the rifled cannon. This example has a dimple and the patent date on the base cup.

Fig. III A-90

Federal

PATTERN: Hotchkiss
DIAMETER: 3.56 inches
GUN: 20-pounder Parrott rifle, 3.67-inch caliber
LENGTH: 10 inches
WEIGHT: 18 pounds, 1 ounce
TYPE: Shell
SABOT: Lead band
FUSING: Percussion, iron anvil cap

Comments: The 20-pounder (3.67-inch caliber) Hotchkiss pattern projectiles, often referred to as long-pattern Hotchkiss projectiles, are rare. It is thought that they were seldom used because they put too much strain on the Parrott rifles.

Fig. III A-91

Federal

PATTERN: Hotchkiss
DIAMETER: 3.57 inches
GUN: 20-pounder Parrott rifle, 3.67-inch caliber
LENGTH: 9 3/4 inches
WEIGHT: 20 pounds, 3 ounces
TYPE: Case shot
SABOT: Lead band
FUSING: Hotchkiss brass fuse plug, paper time fuse

Comments: This is a long-pattern Hotchkiss projectile for the 20-pounder Parrott rifle. This specimen was recovered near the site of the 1864 Battle of Kennesaw Mountain in Georgia.

Fig. III A-92

Federal

PATTERN: Hotchkiss
DIAMETER: 3.74 inches
GUN: 14-pounder James rifle, 3.8-inch caliber
LENGTH: 7 1/8 inches
WEIGHT: 13 pounds, 13 ounces
TYPE: Solid shot
SABOT: Lead band
FUSING: None

Comments: The James projectiles nearly always threw the lead sabot after firing, endangering friendly troops and tending to cause an unstable flight. This led to the replacement of the James projectile with a Hotchkiss pattern (pictured) that could be fired from the existing James rifled cannon. The patent date is cast into the base cup. The fifteen lands and grooves of the James rifling is evident on the lead band. Solid-shot Hotchkiss projectiles are also found in the following field artillery calibers: 2.6-inch, 3-inch, 3.4-inch, 3.5-inch, and in 3.67-inch.

Fig. III A-93

Federal

PATTERN: Hotchkiss
DIAMETER: 3.74 inches
GUN: 14-pounder James rifle, 3.8-inch caliber
LENGTH: 6 7/8 inches
WEIGHT: 14 pounds, 4 ounces
TYPE: Case shot
SABOT: Lead band
FUSING: Hotchkiss brass fuse plug, paper time fuse

Comments: This is one of the Hotchkiss patterns that was used to replace the poorly designed James projectile. It is found in case shot and common shell. The case-shot variety has a more rounded nose than the shell. Specimens have been noted with and without flame grooves. The pictured specimen was recovered from Vicksburg, Mississippi.

Fig. III A-94

Federal

PATTERN: Hotchkiss
DIAMETER: 3.73 inches
GUN: 14-pounder James rifle, 3.8-inch caliber
LENGTH: 7 inches
WEIGHT: 13 pounds, 7 ounces
TYPE: Shell
SABOT: Lead band
FUSING: Percussion, iron anvil cap

Comments: The Hotchkiss projectile pictured has an iron anvil cap percussion fuse that was patented by C.W. Smith and T.D. Stetson on March 25, 1862, patent #34,788. This fuse design was first used at West Point, New York, on its test range for artillery projectiles.

Fig. III A-95

Federal

PATTERN:	Hotchkiss
DIAMETER:	3.73 inches
GUN:	14-pounder James rifle, 3.8-inch caliber
LENGTH:	7 1/8 inches
WEIGHT:	13 pounds, 2 ounces (case-shot balls missing)
TYPE:	Case shot
SABOT:	Lead band
FUSING:	Zinc fuse plug, paper time fuse

Comments: C.W. Smith, G.H. Babcock, B.B. Hotchkiss, and C.A. Hotchkiss were issued patent #38,359 on April 28, 1863, for an improvement on the Hotchkiss case-shot projectile. This consisted of an iron plate that separated the bursting charge and the case-shot chamber. This plate decreased the chances of premature discharge during firing. In order to allow this plate to be inserted easier, the removable nose of the projectile body was developed. As a bonus, this allowed for easier loading of the bursting charge and case-shot balls.

Fig. III A-96

Confederate

PATTERN: James copy, Type I
DIAMETER: 3.21 inches
GUN: 3.3-inch Confederate rifle
LENGTH: 7 3/4 inches
WEIGHT: 11 pounds, 9 ounces
TYPE: Solid shot
SABOT: Lead ring
FUSING: None

Comments: This solid shot was an attempt by the Confederates to copy the James Type I projectile. The specimen has a hollow base that allowed the explosive charge of the cannon to expand the lead ring into the grooves of the rifle. This pattern is extremely rare and is the only copy of the Federal James projectile that the Confederates manufactured. Most fired specimens are found with the lead sabot missing and the base broken. Most specimens have been recovered at Shiloh, Tennessee.

Fig. III A-97

Federal

PATTERN:	James, Type I
DIAMETER:	3.68 inches
GUN:	14-pounder James rifle, 3.8-inch caliber
LENGTH:	7 inches
WEIGHT:	13 pounds, 11 ounces
TYPE:	Solid shot
SABOT:	Lead ring covered with a tin sleeve
FUSING:	None

Comments: This is a non-excavated, unfired James projectile. The lead sabot was covered with a tin sleeve over which canvas was placed. The greased canvas fabric helped lubricate the bore. The purpose of the tin sleeve was to help prevent the lead from fouling the grooves of the rifled cannon. This pattern was patented by Charles T. James on February 26, 1856, patent #14,315 and is classified as Type I due to the date of the patent.

Fig. III A-98

Federal

PATTERN:	James, Type I
DIAMETER:	3.74 inches
GUN:	14-pounder James rifle, 3.8-inch caliber
LENGTH:	7 inches
WEIGHT:	10 pounds, 7 ounces
TYPE:	Shell
SABOT:	Lead ring covered with a tin sleeve (missing)
FUSING:	James percussion

Comments: This specimen illustrates the distinctive ribs of the James projectile. Most of the Type I James projectiles lost their lead sabot after leaving the cannon. When fired over friendly troops, this was found to be unacceptable. Attempts to correct this led to the development of the James Type II projectile. James Type II projectiles were patented on June 10, 1862, patent #35,521. See page 150, figure III A-103, for an example of the Type II.

Fig. III A-99

Federal

PATTERN:	James, Type I
DIAMETER:	4.58 inches
GUN:	12-pounder rifle, 4.62-inch caliber
LENGTH:	8 3/8 inches
WEIGHT:	16 pounds, 14 ounces
TYPE:	Shell
SABOT:	Lead ring covered with a tin sleeve (missing)
FUSING:	James percussion

Comments: The 12-pounder rifle of 4.62-inch caliber saw limited battlefield use. As a result, this caliber of James projectile is uncommon.

Fig. III A-100

Federal

PATTERN: James, Type I
DIAMETER: 3.72 inches
GUN: 14-pounder James rifle, 3.8-inch caliber
LENGTH: 7 3/8 inches
WEIGHT: 11 pounds, 14 ounces
TYPE: Solid shot
SABOT: Lead ring covered with a tin sleeve (missing)
FUSING: None

Comments: This is a variant of the James Type I projectile. It is commonly called a tie-ring James, it is only found in 3.8-inch caliber. The tie-ring solid shot of this variety is rare.

Fig. III A-101

Federal

PATTERN: James, Type I
DIAMETER: 3.69 inches
GUN: 14-pounder James rifle, 3.8-inch caliber
LENGTH: 7 3/8 inches
WEIGHT: 9 pounds, 14 ounces
TYPE: Shell
SABOT: Lead ring covered with a tin sleeve (missing)
FUSING: James percussion, brass

Comments: The pictured specimen is a variant of the James Type I projectile. It is commonly called a tie-ring James due to the visible ring in the base. This ring is felt by some authorities to be used as a means of attaching the cartridge bag. Note the small notches in the ribs best seen on the right view of the projectile. These notches were for wire that was used to help keep the lead sabot attached.

Fig. III A-102

Federal

PATTERN: James, Type II
DIAMETER: 3.75 inches
GUN: 14-pounder James rifle, 3.8-inch caliber
LENGTH: 7 3/4 inches
WEIGHT: 16 pounds, 14 ounces
TYPE: Solid shot
SABOT: Lead with a tin sleeve
FUSING: None

Comments: This James Type II projectile variety was patented on June 10, 1862, patent #35,521. Projectiles of this pattern have been found at Vicksburg, Mississippi. This pattern of James has eight tapering vertical flanges on the projectile body. The lead sabot has been cast in-between these flanges. The exterior of the sabot is covered with tin and canvas to help prevent the lead from fouling the bore of the rifled cannon. See page 150, figure III A-103 for a view of the flanges.

Fig. III A-103

Federal

PATTERN: James, Type II
DIAMETER: 3.73 inches
GUN: 14-pounder James rifle, 3.8-inch caliber
LENGTH: 7 3/4 inches
WEIGHT: 12 pounds
TYPE: Shell
SABOT: Lead with a tin sleeve (missing)
FUSING: James percussion

Comments: James patented his Type II projectile on June 10, 1862, patent #35,521. The projectile base has eight vertical flanges and is tapered in a manner similar to that of a Schenkl. The flanges are wide on the circumference but narrow as they join the shell. The last ordnance contract for the manufacture of James projectiles was on July 28, 1864. See page 24, figure II B-10, for a cross-sectional view. Specimens have been recovered in Vicksburg, Mississippi.

Fig. III A-104

Federal

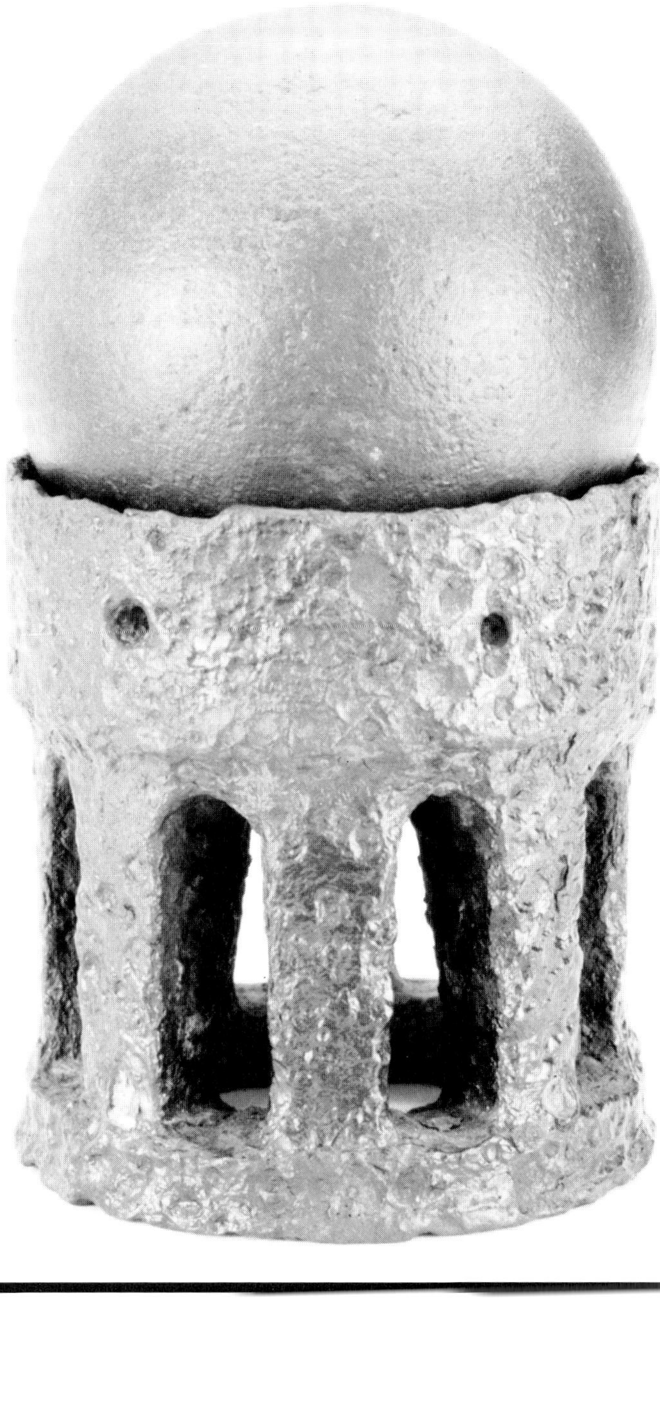

PATTERN: James, Type I
DIAMETER: 3.63 inches
GUN: 6-pounder smoothbore or 3.67-inch caliber rifle
LENGTH: 3 3/4 inches (base only)
WEIGHT: 4 pounds, 5 ounces (base only)
TYPE: Hot shot
SABOT: Lead ring covered with a tin sleeve (missing)
FUSING: None

Comments: This iron base was designed to keep the heated hot shot projectile from making contact with the powder charge in the cannon, which could cause a premature discharge. It was designed to be used in a 3.67-inch rifled cannon or a 3.67-inch smoothbore cannon. Charles T. James patented this projectile, patent #34,950, on April 15, 1862. On the same date, James also obtained patent #34,965 for a similarly constructed base for the 6-pounder Bormann-fused spherical projectile.

Fig. III A-105

Federal

PATTERN:	James, Type I
DIAMETER:	4.45 inches
GUN:	12-pounder rifle and/or smoothbore, 4.62-inch caliber
LENGTH:	4 1/6 inches
WEIGHT:	7 pounds, 14 ounces
TYPE:	Hot shot
SABOT:	Lead ring covered with a tin sleeve (missing)
FUSING:	None

Comments: This very rare pattern of hot shot was found at the Battle of Shiloh, Tennessee, which took place in 1862. The same comments apply as those on page 151, figure III A-104.

Fig. III A-106

Great Britain

PATTERN: Lynall Thomas
DIAMETER: 3.46 inches
GUN: 12-pounder Blakely rifle, 3.5-inch caliber
LENGTH: 7 inches
WEIGHT: 9 pounds, 13 ounces
TYPE: Shell
SABOT: Two metal bands (missing)
FUSING: Wooden fuse plug, paper time fuse

Comments: This projectile is the invention of Lynall Thomas of Great Britain. The center ring or band of iron of the pictured specimen is forward of its original position. The ring was originally centered between two bands, which were most likely made of lead. The wooden shipping plug in the nose is 6 1/2" long. It functioned in a manner similar to that of the Federal Hotchkiss projectile. Upon firing, the base was forced along the cylindrical body, thus expanding the two lead bands into the rifling. This pattern is extremely rare, and most known specimens have been recovered in the vicinity of Charleston, South Carolina.

Fig. III A-107

Confederate

PATTERN: Mullane, sub-pattern I
DIAMETER: 2.16 inches
GUN: Confederate mountain rifle, 2.25-inch caliber
LENGTH: 5 5/8 inches
WEIGHT: 2 pounds, 11 ounces
TYPE: Shell
SABOT: Copper plate
FUSING: Wooden fuse plug, paper time fuse

Comments: This pattern is commonly called the Mullane or Tennessee-sabot shell. The Confederate Patent Office refused a patent for the Mullane projectile on the grounds that the design was anticipated by the patent of Dr. John B. Read. This pattern has a saucer-shaped copper sabot with a wooden dowel pin attached under it to aid in a uniform expansion. A small center bolt held the dowel and sabot in place. The pictured specimen is missing its wooden dowel. Specimens have been recovered from Augusta, Georgia, and near Selma, Alabama.

Fig. III A-108

Confederate

PATTERN:	Mullane, sub-pattern I
DIAMETER:	2.18 inches
GUN:	Confederate mountain rifle, 2.25-inch caliber
LENGTH:	4 5/8 inches
WEIGHT:	2 pounds, 11 ounces
TYPE:	Shell
SABOT:	Copper plate
FUSING:	Copper fuse plug, paper time fuse

Comments: The copper fuse plug shown in this specimen is smaller than the standard Confederate copper fuse plug used in Read projectiles. The central bolt is flush with the sabot. Three iron pins protrude from the base of the projectile through corresponding holes in the copper plate. These iron pins were designed to help prevent the sabot from spinning around the center bolt when fired.

Fig. III A-109

Confederate

PATTERN:	Mullane, sub-pattern I
DIAMETER:	2.94 inches
GUN:	3-inch wrought iron (ordnance) rifle
LENGTH:	7 15/16 inches
WEIGHT:	7 pounds, 3 ounces
TYPE:	Shell
SABOT:	Copper plate
FUSING:	Wooden fuse plug, paper time fuse

Comments: This is the most common Mullane variety recovered. It is often found with the copper sabot deformed or missing. An article written in 1866 by Confederate General E.P. Alexander found in the *Southern Historical Society Papers* states that the Mullane shell ". . . failed, however, about three times out of four from breaking its connection with the copper sabot, and it very frequently exploded in the gun; while of those which flew correctly, not one-fourth exploded at all."

Fig. III A-110

Confederate

PATTERN:	Mullane, sub-pattern I
DIAMETER:	2.94 inches
GUN:	3-inch wrought iron (ordnance) rifle
LENGTH:	7 3/4 inches
WEIGHT:	7 pounds, 3 ounces
TYPE:	Shell
SABOT:	Copper plate
FUSING:	Copper fuse plug, paper time fuse

Comments: The typical Mullane projectile in this caliber is found with a wooden fuse plug for a paper time fuse. This specimen has a copper fuse plug and was found near the battlefield of Kennesaw Mountain, Georgia, which took place in 1864. Attempts were made to ensure the ignition of the fuse by filing notches in the copper sabot to allow the flame of the discharge to pass. This met with limited success. Note the one flame groove cut out in this pictured projectile. Approximately one-fourth of the Mullane projectiles that were fired exploded.

Fig. III A-111

Confederate

PATTERN:	Mullane, sub-pattern I
DIAMETER:	2.94 inches
GUN:	3-inch wrought iron (ordnance) rifle
LENGTH:	8 3/8 inches
WEIGHT:	7 pounds, 4 ounces
TYPE:	Case shot
SABOT:	Copper plate
FUSING:	Copper fuse plug, paper time fuse

Comments: The Confederates substituted iron case-shot balls for lead due to the shortage of lead needed for small arms ammunition. The Confederates could not drill into the iron balls to form the bursting cavity without great difficulty. The bursting cavity was formed by a dowel inserted into the fuse opening. The iron balls were inserted through the hole on the projectile's side, and the hot, liquefied matrix was poured into the side-loading hole. After the matrix cooled, the dowel was removed, a threaded lead plug was screwed into the loading hole, powder installed, and the fuse adapter was screwed in. Side-loader rifled projectiles are rare.

Fig. III A-112

Confederate

PATTERN:	Mullane, sub-pattern I
DIAMETER:	3.45 inches
GUN:	12-pounder Blakely rifle, 3.5-inch caliber
LENGTH:	7 1/4 inches
WEIGHT:	11 pounds, 5 ounces
TYPE:	Shell
SABOT:	Copper plate
FUSING:	Wooden fuse plug, paper time fuse

Comments: The sabot on this pictured projectile has three iron pins that protrude through the copper plate and one center bolt flush with the sabot plate. The Confederates manufactured this pattern for the British 3.5-inch Blakely rifle in order to keep up with the demand for ammunition; there was only a limited supply of Britten projectiles imported from Great Britain. Note the seven lands and grooves of the Blakely rifle on the copper sabot.

Fig. III A-113

Read vs. Parrott

Comments: The authors feel that a very important area of potential confusion exists in the ability to tell the difference between a smooth-sided Confederate Read and a Federal Parrott projectile. Both projectiles are pictured here for comparison. On the reader's left is a Read, which can be identified by the lathe dimple in the base of the projectile body, a thinner wrought iron sabot, and lathing marks that run the entire length of the cylindrical body. Although difficult to see in this particular photograph, the slope, the angle, and the space between the wrought iron sabot and the bottom of the cylindrical body is slightly different in appearance than that of the Parrott projectile on the reader's right. The Parrott projectile exhibits a large chiseled area on the base where the gate waste was removed after casting. Also note there is no evidence of a lathe dimple, since the cylindrical body was not machine-turned. The wrought iron cup is thicker and pre-stamped with corresponding rifling projections in nearly all Parrott projectiles; only rarely do Read projectiles exhibit similar pre-stamped projections.

Differences can also be seen when comparing the commonly found fuses. The Read projectile is most often found with a wooden fuse plug or a copper fuse plug. The Parrott projectile is commonly found with a white metal fuse plug made of zinc.

We hope the above information helps the collector to determine the correct pattern of a questionable projectile. There is, however, no substitute for hands-on experience.

See page 182, figure III A-135, for a Read projectile, and see page 165, figure III A-118, for a Parrott projectile. For more information concerning Read vs. Parrott, see page 181.

Fig. III A-114

Federal

PATTERN: Parrott, sub-pattern I
DIAMETER: 2.86 inches
GUN: 10-pounder Parrott rifle, 2.9-inch caliber
LENGTH: 6 1/2 inches
WEIGHT: 9 pounds, 9 ounces
TYPE: Solid shot
SABOT: Wrought iron cup
FUSING: None

Comments: Robert P. Parrott patented this projectile on August 20, 1861, patent #33,100. This was an attempt to correct the problem that John B. Read had with his patented thin wrought iron cup projectile that was patented on October 28, 1856, patent #15,999. Parrott's improvement to Read's patent was a thicker wrought iron cup or ring that was manufactured with three corresponding rifling projections stamped into the sabot. This helped to prevent the breakage of the wrought iron cup during expansion. This specimen is commonly called a Parrott flat-nose bolt. They are also found in 20-pounder (3.67-inch caliber) size. See page 181, figure III A-134, for additional comments.

Fig. III A-115

Federal

PATTERN:	Parrott, sub-pattern III
DIAMETER:	2.86 inches
GUN:	10-pounder Parrott rifle, 2.9-inch caliber
LENGTH:	6 9/16 inches
WEIGHT:	10 pounds, 4 ounces
TYPE:	Solid shot
SABOT:	Brass ring
FUSING:	None

Comments: This Parrott flat-nose bolt was found in Vicksburg, Mississippi. The Parrott rifles of 3-inch caliber were not delivered to the Federal army until February of 1864. The Siege of Vicksburg ended on July 4, 1863, so all 10-pounder Parrott projectiles with brass or iron sabots found on battlefields that date prior to February 1864 were fired from the 10-pounder Parrott rifles of 2.9-inch caliber.

Fig. III A-116

Federal

PATTERN:	Parrott, sub-pattern II
DIAMETER:	2.94 inches
GUN:	3-inch Parrott rifle
LENGTH:	6 3/8 inches
WEIGHT:	10 pounds, 2 ounces
TYPE:	Solid shot
SABOT:	Brass ring
FUSING:	None

Comments: This rare high-band Parrott projectile has a chilled nose for punching through fortifications, earthworks, or other structures. Parrott patented this sabot ring on August 20, 1861, patent #33,099. The iron projectile body was cast and, while still hot, forced through a sizing die (note the horizontal lines on the body). After the projectile body was taken from the mould, the brass ring sabot was cast on the projectile and machined to tolerance. The base was then drilled with four holes between the base and the brass ring to aid in the expansion of the sabot. This variety of sabot frequently did not enter the grooves of the rifle. The side of the projectile body pictured is stamped "NCR" and "PATENTED 1861."

Fig. III A-117

Federal

PATTERN:	Parrott, sub-pattern III
DIAMETER:	2.87 inches
GUN:	10-pounder Parrott rifle, 2.9-inch caliber
LENGTH:	6 13/16 inches
WEIGHT:	9 pounds, 11 ounces
TYPE:	Solid shot
SABOT:	Brass ring
FUSING:	None

Comments: The majority of battlefield recoveries of this variety of 10-pounder (2.9-inch caliber) Parrott chilled-nose bolt projectile have been at Vicksburg, Mississippi. The purpose of this bolt was to punch through fortifications, buildings, and other structures. This style of projectile is also found in a 20-pounder (3.67-inch caliber) size.

Fig. III A-118

Federal

PATTERN:	Parrott, sub-pattern I
DIAMETER:	2.86 inches
GUN:	10-pounder Parrott rifle, 2.9-inch caliber
LENGTH:	8 7/8 inches
WEIGHT:	8 pounds, 15 ounces
TYPE:	Shell
SABOT:	Wrought iron cup
FUSING:	Parrott percussion, zinc

Comments: All Federal Parrott projectiles with wrought iron cup sabots were for use in the 10-pounder Parrott rifle, 2.9-inch caliber, and are designated sub-pattern I. Robert P. Parrott patented this projectile on August 20, 1861, patent #33,100. The wrought iron cup was cast first and placed into the projectile mould in such a manner that, after casting, it was firmly embedded in the metal body. It was then placed in a metal cylinder and pre-stamped with the rifling grooves during the manufacturing process. The fuse in this specimen is the first model of Parrott percussion that was developed in West Point, New York.

Fig. III A-119

Federal

PATTERN:	Parrott, sub-pattern I
DIAMETER:	2.86 inches
GUN:	10-pounder Parrott rifle, 2.9-inch caliber
LENGTH:	8 1/2 inches
WEIGHT:	11 pounds, 7 ounces
TYPE:	Case shot
SABOT:	Wrought iron cup
FUSING:	Bormann time fuse

Comments: The fusing of a Parrott projectile with the 5 1/4-second Bormann time fuse is highly unusual. The burning time made this fuse inadequate for long-range rifled cannon. See page 26, figure II B-12, for a cross-sectional view. Bormann-fused Parrott projectiles were manufactured in limited numbers. Bormann fuses are also found in Confederate 10-pounder smooth-sided Read projectiles.

Fig. III A-120

Federal

PATTERN:	Parrott, sub-pattern II
DIAMETER:	2.94 inches
GUN:	3-inch Parrott rifle
LENGTH:	8 7/8 inches
WEIGHT:	11 pounds, 11 ounces
TYPE:	Case shot
SABOT:	Brass ring
FUSING:	Parrott zinc fuse plug, paper time fuse

Comments: Robert P. Parrott patented this projectile on August 20, 1861, patent #33,099. The brass sabot ring was cast around the base of the projectile, which had six indentations. The purpose of these indentations was to help keep the sabot from slipping. All of the high-band pattern Parrott projectiles with this sabot variety are classified as sub-pattern II. Most often the sabot would not enter the grooves of the rifle cannon upon firing due to the close tolerance between the sabot and the shell base. This resulted in the chipping of the iron base of the projectile. Many recoveries have had the sabot sawed to correct the expansion problems. This pattern is also found in 10-pounder Parrott (2.9-inch caliber) and 20-pounder (3.67-inch caliber). See page 25, figure II B-11, for a cross-sectional example of the projectile's interior.

Fig. III A-121

Federal

PATTERN: Parrott, sub-pattern III
DIAMETER: 2.86 inches
GUN: 10-pounder Parrott rifle, 2.9-inch caliber
LENGTH: 8 9/16 inches
WEIGHT: 10 pounds, 2 ounces
TYPE: Case shot
SABOT: Brass ring
FUSING: Parrott zinc fuse plug, paper time fuse

Comments: In order to correct the chipping problem of the sub-pattern II sabots, a sabot was developed with five internal flanges. However, this caused the bottom of the sabot to be flush with the base of the shell which, in combination with the extreme thickness of the sabot, would not allow the explosive charge to force the sabot into the grooves of the gun tube. Occasionally the shells failed to take the rifling of the bore, thereby decreasing the range and accuracy. The projectile body was cast and sized with a die while hot. The brass ring was cast on the projectile base after this procedure. This projectile is also found in a 20-pounder size (3.67-inch caliber). See page 25, figure II B-11, for a cross-sectional view.

Fig. III A-122

Federal

PATTERN:	Parrott, sub-pattern III
DIAMETER:	3.58 inches
GUN:	20-pounder Parrott rifle, 3.67-inch caliber
LENGTH:	7 13/16 inches
WEIGHT:	18 pounds, 10 ounces
TYPE:	Solid shot
SABOT:	Brass ring
FUSING:	None

Comments: This is often referred to as a 20-pounder (3.67-inch caliber) flat-nose bolt. It is also found in a 10-pounder size (2.9-inch caliber). Note the nice rifling of the Parrott rifle on the brass sabot. This caliber is uncommon in the flat-nose bolt variety.

Fig. III A-123

Federal

PATTERN: Parrott, sub-pattern III
DIAMETER: 3.63 inches
GUN: 20-pounder Parrott rifle, 3.67-inch caliber
LENGTH: 8 7/8 inches
WEIGHT: 19 pounds, 14 ounces
TYPE: Solid shot
SABOT: Brass ring
FUSING: None

Comments: This projectile is commonly referred to as a 20-pounder (3.67-inch caliber) Parrott chilled-nose bolt. It was designed primarily for destroying fortifications and was not often used as an anti-personnel projectile. It is also found in a 10-pounder (2.9-inch caliber) size. Note the expansion of the brass sabot.

Fig. III A-124

Federal

PATTERN:	Parrott, sub-pattern I
DIAMETER:	3.63 inches
GUN:	20-pounder Parrott rifle, 3.67-inch caliber
LENGTH:	9 1/4 inches
WEIGHT:	17 pounds, 15 ounces
TYPE:	Case shot
SABOT:	Wrought iron cup
FUSING:	Parrott zinc fuse plug, paper time fuse

Comments: Typically, the short variety of the 20-pounder (3.67-inch caliber) Parrott shell was filled with lead case-shot balls. The projectile was shorter in order to keep the weight down due to the added case-shot material.

Fig. III A-125

Federal

PATTERN:	Parrott, sub-pattern III
DIAMETER:	3.63 inches
GUN:	20-pounder Parrott rifle, 3.67-inch caliber
LENGTH:	10 1/8 inches
WEIGHT:	18 pounds, 1 ounce
TYPE:	Shell
SABOT:	Brass ring
FUSING:	Parrott zinc fuse plug, paper time fuse

Comments: This is an example of the long-pattern 20-pounder (3.67-inch caliber) Parrott projectile. Typically, the long-pattern Parrott projectiles are common shell and not case-shot. This variety is also found with a wrought iron sabot.

Fig. III A-126

Federal

PATTERN: Pattison
DIAMETER: 2.31 inches
GUN: 6-pounder Pattison rifle, 2.32-inch caliber
LENGTH: 6 11/16 inches
WEIGHT: 5 pounds, 14 ounces
TYPE: Solid shot
SABOT: None
FUSING: None

Comments: An 1861 issue of *Scientific American* described this projectile and matching rifle as being manufactured by a company in Troy, New York. It has five smooth-shaped flanges and five vertical grooves on the projectile body. It had a leather sleeve around the lower third of the projectile body to help prevent the propellant gas from escaping.

Fig. III A-127

Great Britain

PATTERN:	Preston
DIAMETER:	3.42 inches across flats
GUN:	12-pounder Blakely rifle, 3.5-inch caliber
LENGTH:	7 11/16 inches
WEIGHT:	11 pounds, 11 ounces
TYPE:	Shell
SABOT:	None
FUSING:	Britten percussion, brass

Comments: This pattern was used only by the Confederate forces and even then in extremely limited quantities. This projectile has a flat base and six vertical inclined planes along the sides to allow it to fit the Blakely rifle's sawtooth rifling. The poor design of the Britten fuse caused many shells to be rendered harmless upon firing. This projectile is also found in a 18-pounder (4-inch caliber) size. The majority of the battlefield recovered 12-pounder (3.5-inch caliber) Preston projectiles have come from Port Hudson, Louisiana.

Fig. III A-128

Confederate

PATTERN: Read
DIAMETER: 2.45 inches
GUN: 2.5-inch Blakely rifle
LENGTH: 6 7/8 inches
WEIGHT: 5 pounds, 12 ounces
TYPE: Shell
SABOT: Copper ring
FUSING: Wooden fuse plug, paper time fuse

Comments: Notice the six sawtooth rifling marks from the 2.5-inch Blakely rifle on the sabot. Specimens have been recovered from Atlanta Campaign sites, and some have been found to have segmented interiors. The 2.5-inch Britten projectile shown on page 98, figure III A-51, was also used in the 2.5-inch Blakely rifle.

Fig. III A-129

Confederate

PATTERN: Read
DIAMETER: 2.57 inches
GUN: 2.6-inch rifle
LENGTH: 7 1/2 inches
WEIGHT: 6 pounds, 6 ounces
TYPE: Shell
SABOT: Copper ring
FUSING: Wooden fuse plug, Girardey percussion

Comments: This unfired specimen was found with a Girardey percussion fuse. The base is slightly rounded and has no evidence of a lathe dimple, although it was turned on a lathe. Note the crude lathe grooves on the body and sabot of the projectile. Most specimens of this pattern have been recovered from South Carolina.

Fig. III A-130

Confederate

PATTERN:	Read
DIAMETER:	2.94 inches
GUN:	3-inch wrought iron (ordnance) rifle
LENGTH:	6 1/4 inches
WEIGHT:	8 pounds, 1 ounces
TYPE:	Solid shot
SABOT:	Copper ring
FUSING:	None

Comments: The 3-inch Read bolt pictured here is the most common recovered pattern. Most specimens have been recovered from Mississippi. Note that the copper sabot is thick and unevenly cast. A few projectiles of this type have been found at Resaca, Georgia, near Atlanta.

Fig. III A-131

Confederate

PATTERN: Read
DIAMETER: 2.94 inches
GUN: 3-inch Confederate rifle
LENGTH: 6 1/4 inches
WEIGHT: 8 pounds
TYPE: Solid shot
SABOT: Copper ring
FUSING: None

Comments: This projectile's nose is more pointed, and the copper sabot is taller and thinner than the specimen shown on page 177, figure III A-130. Note the nice twelve lands and grooves of the rifling on the copper sabot. This specimen was found at Vicksburg, Mississippi.

Fig. III A-132

Confederate

PATTERN: Read
DIAMETER: 2.94 inches
GUN: 3-inch Parrott rifle
LENGTH: 6 3/8 inches
WEIGHT: 9 pounds, 15 ounces
TYPE: Solid shot
SABOT: Brass ring
FUSING: None

Comments: This is a crude copy of the 3-inch Federal Parrott chilled-nose bolt. The brass sabot is shorter than its Federal counterpart and has no indentations. See page 163, figure III A-116, for comparison.

Fig. III A-133

Confederate

PATTERN:	Read
DIAMETER:	2.94 inches
GUN:	3-inch Confederate rifle
LENGTH:	7 3/16 inches
WEIGHT:	10 pounds, 8 ounces
TYPE:	Solid shot
SABOT:	Copper ring
FUSING:	None

Comments: This specimen is rare and often found with the thin, tall sabot bent over or missing. The sabot extends 1/2" below the flat iron base. Notice the excellent left-hand twist of the twelve lands and grooves of the 3-inch Confederate rifling on the copper sabot.

Fig. III A-134

Pre-war Federal

PATTERN:	Read-Parrott
DIAMETER:	2.87 inches
GUN:	10-pounder Parrott rifle, 2.9-inch caliber
LENGTH:	8 3/4 inches
WEIGHT:	8 pounds, 14 ounces
TYPE:	Shell
SABOT:	Wrought iron cup
FUSING:	Wooden fuse plug, paper time fuse

Comments: Robert Parrott and John Read were both projectile inventors before the Civil War. Read patented his wrought iron ring sabot projectiles on October 28, 1856, patent #15,999. Parrott purchased the rights to manufacture Read's projectiles before 1861, and a royalty was to be paid to Read. On August 20, 1861, with patent #33,100, Parrott patented an improvement to Read's projectiles. Parrott's patent stated, in part: "This invention consists in an improvement upon the elongated projectile for which letters Patent of the United States were issued on the 28th day of October, 1856, to John B. Read . . . . the cups [Read sabots] became so weak as to be liable to break away from the body of the projectile . . . . to prevent this, I make the said cup, more especially at its edges, of greater thickness, and to insure its proper entrance into the grooves I swage or otherwise form the said cup before the insertion of the projectile in the gun in such a manner that in loading it it will enter into the grooves in such a manner that it will not interfere with the free loading of the gun, but that it will be driven completely into the grooves and caused to fit the grooves and lands by the force of the explosion of the charge of powder without any danger of its being broken."

According to a letter written by Parrott, this projectile pattern was tested at the Washington Arsenal under the command of Bridadier General George D. Ramsay in June 1861 was used exclusively by Ricketts' Battery at First Manassas. The projectile shown here, which was recovered from the site of the Battle of First Manassas, is exactly like that drawn in Parrott's patent. Although it is of Federal manufacture, it falls under Read's patent of October 28, 1856; Parrott only made an improvement to Read's wrought iron cup (sabot). Therefore, all wrought iron sabot Read or Parrott projectiles should properly be classified as Read if of Confederate manufacture and Parrott if of Federal manufacture. Parrott later patented a brass ring sabot which should properly be referred to by his name only. The later Parrott patterns will have been threaded for fuses and will have brass sabots.

However, to lessen confusion, the authors have elected to continue the generally accepted terminology—i.e., Parrott.

Fig. III A-135

Confederate

PATTERN: Read
DIAMETER: 2.87 inches
GUN: 10-pounder Parrott rifle, 2.9-inch caliber
LENGTH: 9 inches
WEIGHT: 9 pounds
TYPE: Shell
SABOT: Wrought iron cup
FUSING: Wooden fuse plug, paper time fuse

Comments: Stamped on both sides of the projectile body are the letters "AR." This is possibly the manufacturing stamp of A.J. Rahm, who was proprietor of Eagle Machine Works in Richmond, Virginia. His company manufactured artillery, artillery carriages, ammunition, and tools, and outfitted gunboats during the period between 1862 and 1864. Several specimens with the "AR" stamp have been recovered from sites in Virginia and South Carolina.

Fig. III A-136

Confederate

PATTERN: Read
DIAMETER: 2.85 inches
GUN: 10-pounder Parrott rifle, 2.9-inch caliber
LENGTH: 8 3/8 inches
WEIGHT: 8 pounds, 2 ounces
TYPE: Shell
SABOT: Wrought iron cup
FUSING: Wooden fuse plug, paper time fuse

Comments: This unfired specimen has three pre-stamped rifling flanges in the wrought iron cup, which were designed to correspond to the cannon's rifling. This is uncommon in the Read pattern projectiles. Note that there is a lathe dimple in the base. The rebated section above the sabot is 3/4" wide and was most likely intended for a lubricating material.

Fig. III A-137

Confederate

PATTERN: Read
DIAMETER: 2.85 inches
GUN: 10-pounder Parrott rifle, 2.9-inch caliber
LENGTH: 9 1/4 inches
WEIGHT: 9 pounds, 15 ounces
TYPE: Shell
SABOT: Wrought iron cup
FUSING: Copper fuse plug, paper time fuse

Comments: The presence of the copper fuse plug indicates that this is a later production of the smooth-sided Read projectile. The smooth early-war appearance continued on Confederate Read projectiles until the difficulties in production led to the development of the bourrelet. The surface of the projectile parallel to the sides of the bore had to be machined to the required dimension to allow for windage. The bourrelet was developed in order to speed up the manufacturing time and save on tooling wear.

Fig. III A-138

Confederate

PATTERN: Read
DIAMETER: 2.86 inches
GUN: 10-pounder Parrott rifle, 2.9-inch caliber
LENGTH: 8 3/4 inches
WEIGHT: 8 pounds, 6 ounces
TYPE: Shell
SABOT: Wrought iron cup
FUSING: Copper fuse plug, paper time fuse

Comments: This projectile has two bourrelets, or bearing surfaces, with the upper being wider than the lower. The bottom bourrelet has been machined down to the diameter of the projectile body. The base is slightly rounded with a lathe dimple and the sabot is made of thin wrought iron. This is an example of a typical Virginia Read most likely manufactured at Tredegar Iron Works in Richmond, Virginia. See page 27, figure II B-13, for a representative cross-sectional view.

Fig. III A-139

Confederate

PATTERN: Read
DIAMETER: 2.85 inches
GUN: 10-pounder Parrott rifle, 2.9-inch caliber
LENGTH: 8 1/2 inches
WEIGHT: 10 pounds, 1 ounce
TYPE: Case shot
SABOT: Wrought iron cup
FUSING: Copper fuse plug, paper time fuse

Comments: Note the chipping around the projectile base, which was due to the poor iron and sabot design used by the Confederates. R.P. Parrott patented an improvement to Read's sabot problem, patent #33,100, on August 20, 1861. Located near the base is a rebated section similar to the projectile on page 183, figure III A-136. It has iron case-shot balls that were inserted where the lead plug is located. This pattern of Read projectile is scarce. This specimen was recovered from the Siege of Vicksburg, Mississippi. See page 158 for a description of the case-shot construction.

Fig. III A-140

Confederate

PATTERN:	Read
DIAMETER:	2.85 inches
GUN:	10-pounder Parrott rifle, 2.9-inch caliber
LENGTH:	8 5/8 inches
WEIGHT:	10 pounds, 2 ounces
TYPE:	Shell
SABOT:	Copper ring
FUSING:	Wooden fuse plug, paper time fuse

Comments: This is an example of a Read pattern with a high, thin copper sabot. Most specimens have been found on battlefields near Atlanta, Georgia. The projectile base has a lathe dimple and is slightly rounded. This style of Read was manufactured at the Atlanta Arsenal.

Fig. III A-141

Confederate

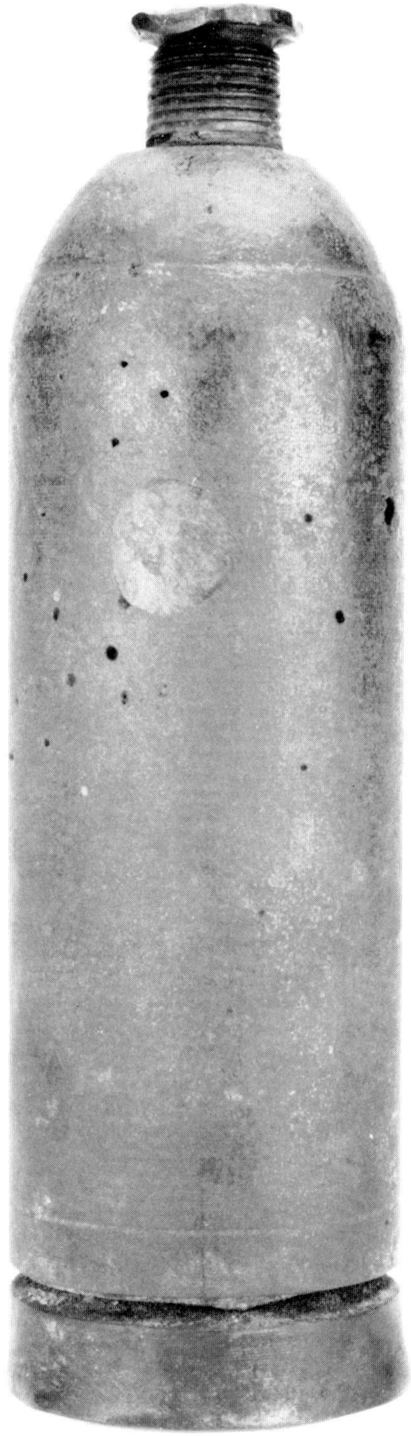

PATTERN: Read
DIAMETER: 2.86 inches
GUN: 10-pounder Parrott rifle, 2.9-inch caliber
LENGTH: 8 7/16 inches
WEIGHT: 10 pounds, 7 ounces
TYPE: Case shot
SABOT: Copper ring
FUSING: Copper fuse plug, paper time fuse

Comments: This pattern has iron case-shot balls and a lead plug sealing the side-loading hole. The copper sabot projects below the iron base. The projectile's cylindrical body is slightly larger than its nose. A similar specimen is noted that has an iron sabot. The threaded fuse plug was hastily cross-threaded into the nose, causing it to extend farther from the nose than usual. See page 158 for a description of how case shot was installed.

Fig. III A-142

Confederate

PATTERN: Read
DIAMETER: 2.93 inches
GUN: 3-inch Parrott rifle
LENGTH: 7 5/8 inches
WEIGHT: 8 pounds
TYPE: Shell
SABOT: Copper ring
FUSING: Wooden fuse plug, paper time fuse

Comments: This projectile is often referred to as a Trans-Mississippi Read due to the fact that all examples of this pattern have been recovered from Western Theatre sites. Three flame grooves are cut out of the thin copper sabot. It is also found in 2.25-inch, 3.3-inch, and 3.5-inch calibers.

Fig. III A-143

Confederate

PATTERN: Read
DIAMETER: 2.86 inches
GUN: 10-pounder Parrott rifle, 2.9-inch caliber
LENGTH: 8 7/8 inches
WEIGHT: 9 pounds, 6 ounces
TYPE: Case shot
SABOT: Copper ring
FUSING: Copper fuse plug, paper time fuse

Comments: This unfired specimen has one flame groove cut out of the high copper sabot. It is case shot and was loaded through the fuse opening. Note the packing toe in the fuse plug. The packing toe protected the powder charge from foreign matter and was removed prior to the insertion of a paper time fuse.

Fig. III A-144

Confederate

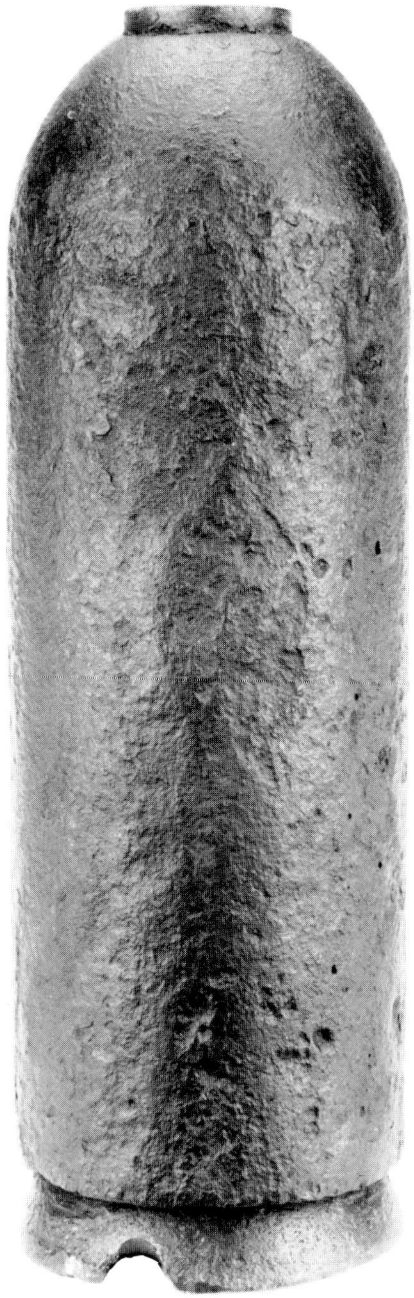

PATTERN:	Read
DIAMETER:	2.84 inches
GUN:	10-pounder Parrott rifle, 2.9-inch caliber
LENGTH:	8 1/2 inches
WEIGHT:	9 pounds, 8 ounces
TYPE:	Shell
SABOT:	Copper ring
FUSING:	Copper fuse plug, paper time fuse

Comments: This pattern is usually found in Atlanta Campaign sites. Projectiles of this style were manufactured at the Atlanta Arsenal. The base is rounded and has a very thick, brittle copper sabot with one flame groove notch cut into it. Fired specimens usually have most of the sabot missing.

Fig. III A-145

Confederate

PATTERN: Read
DIAMETER: 2.94 inches
GUN: 3-inch wrought iron (ordnance) rifle
LENGTH: 7 5/16 inches
WEIGHT: 8 pounds
TYPE: Shell
SABOT: Copper ring
FUSING: Wooden fuse plug, paper time fuse

Comments: This Read pattern was developed out of the necessity to save man hours and tooling costs. Instead of having to turn the entire length of the projectile on a lathe, only the two bourrelet surfaces had to be machined. This is an early bourrelet pattern with a wooden fuse plug. Note the distinct mould seam running the entire length of the projectile body. The sabot shows excellent rifling from the 3-inch rifle.

Fig. III A-146

Confederate

PATTERN: Read
DIAMETER: 2.94 inches
GUN: 3-inch Parrott rifle
LENGTH: 7 3/16 inches
WEIGHT: 7 pounds, 12 ounces
TYPE: Shell
SABOT: Copper ring
FUSING: Copper fuse plug, paper time fuse

Comments: This is the most common variety of Read projectile recovered. A lathe dimple is found in the center of the iron base. This pattern of Read was designed to be fired from the 3-inch wrought iron (ordnance) rifle and not the Parrott rifle. The pictured specimen was fired from a 3-inch Parrott rifle. The same comments apply as those on page 192, figure III A-145.

Fig. III A-147

Confederate

PATTERN:	Read
DIAMETER:	2.94 inches
GUN:	3-inch wrought iron (ordnance) rifle
LENGTH:	7 3/16 inches
WEIGHT:	Not applicable due to chipped iron base
TYPE:	Shell
SABOT:	Copper ring
FUSING:	Copper fuse plug, paper time fuse

Comments: The poor quality of the iron in the Confederate-manufactured projectiles is evident in this specimen. When the sabot was forced into the grooves of the rifle, the expansion of the sabot often broke the iron around the base.

Fig. III A-148

Confederate

PATTERN: Read
DIAMETER: 2.94 inches
GUN: 3-inch Confederate rifle
LENGTH: 7 1/2 inches
WEIGHT: 8 pounds, 4 ounces
TYPE: Shell
SABOT: Copper ring
FUSING: Copper fuse plug, paper time fuse

Comments: This high-band Read has excellent rifling from the twelve lands and grooves of the 3-inch Confederate rifle. Projectiles of this pattern are found in Atlanta Campaign sites. The two bourrelets on this variety are narrower than those typically found. The pictured specimen was found at a site relating to the Atlanta Campaign.

Fig. III A-149

Confederate

PATTERN: Read
DIAMETER: 2.94 inches
GUN: 3-inch wrought iron (ordnance) rifle
LENGTH: 7 3/4 inches
WEIGHT: 7 pounds, 14 ounces
TYPE: Shell
SABOT: Copper ring
FUSING: Confederate percussion

Comments: This specimen has a rare Confederate percussion fuse with a screwdriver slot. Most of the specimens of this style were fitted with time fuse plugs. Note that the lathe dog is still attached. Usually the lathe dog was broken off the projectile before it left the arsenal.

Fig. III A-150

Confederate

PATTERN:	Read
DIAMETER:	2.96 inches
GUN:	3-inch wrought iron (ordnance) rifle
LENGTH:	7 5/16 inches
WEIGHT:	8 pounds, 14 ounces
TYPE:	Case shot
SABOT:	Copper ring
FUSING:	Copper fuse plug, paper time fuse

Comments: Because the Confederates desperately needed lead for small arms ammunition, they used iron rather than lead for case-shot material. This presented a difficult situation, as they could not drill into the iron balls to form the bursting cavity without damaging the drill bits or breaking the matrix loose. The bursting cavity was formed by inserting a dowel into the fuse opening. The iron balls were inserted through the hole on the projectile's side, and the hot, liquefied matrix was poured into the side-loading hole. After the matrix cooled, the dowel was removed, a lead plug screwed into the loading hole, powder installed, and the fuse adapter screwed into the nose of the projectile. Side-loading rifled projectiles are rare. The projectile pictured has three frame grooves cut into the sabot, a lathe dog (or lug) near the top of the nose, a lathe dimple in the base, and was loaded with iron case-shot balls.

Fig. III A-151

Confederate

PATTERN:	Read
DIAMETER:	2.94 inches
GUN:	3-inch wrought iron (ordnance) rifle
LENGTH:	7 1/8 inches
WEIGHT:	8 pounds, 4 ounces
TYPE:	Shell
SABOT:	Copper ring
FUSING:	Wooden fuse plug, paper time fuse

Comments: The lower bourrelet of this Read projectile is wider than the upper one. This style is often referred to as a high-band Read due to the taller-than-usual sabot. Note the casting imperfections in the copper sabot. The groove between the sabot and the projectile body was cut by the lathe tool. This space was for a lubrication and safety expansion groove. This pattern of projectile has been recovered in South Carolina.

Fig. III A-152

Confederate

PATTERN: Read
DIAMETER: 2.94 inches
GUN: 3-inch wrought iron (ordnance) rifle
LENGTH: 7 1/4 inches
WEIGHT: 8 pounds, 2 ounces
TYPE: Shell
SABOT: Copper ring
FUSING: Wooden fuse plug, paper time fuse

Comments: This is an unfired smooth-sided Read with a high copper sabot cast flush with the side of the projectile body. Notice the crude mould seam around the upper third of the projectile; Confederate mould seams usually ran longitudinally. This pattern of Read projectile is scarce. It is similar to the specimen on page 198, figure III A-151, except that it has no bourrelets.

Fig. III A-153

Confederate

PATTERN: Read
DIAMETER: 3.23 inches
GUN: 3.3-inch Confederate rifle
LENGTH: 5 3/4 inches
WEIGHT: 8 pounds
TYPE: Solid shot
SABOT: Copper ring
FUSING: None

Comments: This Read bolt was manufactured at Selma Arsenal, Selma, Alabama. It was intended to be used in the 3.3-inch Confederate cannon. Note the two wide bourrelets and the high copper sabot. The variety of Read has been recovered from Selma, Alabama; Milledgeville, Georgia; and Shiloh, Tennessee.

Fig. III A-154

Confederate

PATTERN:	Read
DIAMETER:	3.23 inches
GUN:	3.3-inch Confederate rifle
LENGTH:	7 inches
WEIGHT:	8 pounds, 7 ounces
TYPE:	Shell
SABOT:	Copper ring
FUSING:	Copper fuse plug, paper time fuse

Comments: *The Official Records,* Series I, Volume XLIV, report number 90 of Colonel William Hawley, 3rd Wisconsin Infantry, states that on or around November 22, 1864, Hawley's troops threw 170 boxes of fixed artillery ammunition into the Oconee River in Milledgeville, Georgia (the capital of Georgia during the war), and they also destroyed the weapons of warfare that the Confederates had stored. Projectiles similar to the specimen shown here have been recovered from Milledgeville, Georgia. This rare unfired Read projectile has a high copper sabot with one flame groove cut into it. It was manufactured at Selma Arsenal in Selma, Alabama.

Fig. III A-155

Confederate

PATTERN: Read
DIAMETER: 3.22 inches
GUN: 3.3-inch Confederate rifle
LENGTH: 6 3/4 inches
WEIGHT: 7 pounds, 9 ounces
TYPE: Shell
SABOT: Copper ring
FUSING: Confederate percussion, copper anvil cap

Comments: This Read projectile was manufactured at the Selma Arsenal in Selma, Alabama. Note the typical high-band sabot usually found on this Read pattern. Specimens have been recovered in Milledgeville, Georgia, and at Selma, Alabama.

Fig. III A-156

Confederate

PATTERN:	Read
DIAMETER:	3.34 inches
GUN:	3.4-inch rifle
LENGTH:	6 1/4 inches
WEIGHT:	8 pounds, 1 ounce
TYPE:	Shell
SABOT:	Wrought iron cup
FUSING:	Wooden fuse plug, paper time fuse

Comments: Note the excellent lathe grooves running the entire length of the cylindrical projectile body. The wrought iron sabot has the impression of three lands and grooves similar to those of a Parrott rifle.

Fig. III A-157

Confederate

PATTERN: Read
DIAMETER: 3.43 inches
GUN: 12-pounder Blakely rifle, 3.5-inch caliber
LENGTH: 7 7/8 inches
WEIGHT: 11 pounds, 7 ounces
TYPE: Shell (segmented)
SABOT: Copper ring
FUSING: Copper fuse plug, paper time fuse

Comments: This unfired specimen was found in a Confederate fort in High Bridge, Virginia. Several hundred projectiles of various patterns were found in this cache. The shortage of imported British ammunition for the 3.5-inch calibre Blakely rifles led to the manufacture of this variety of Read projectiles.

Fig. III A-158

Confederate

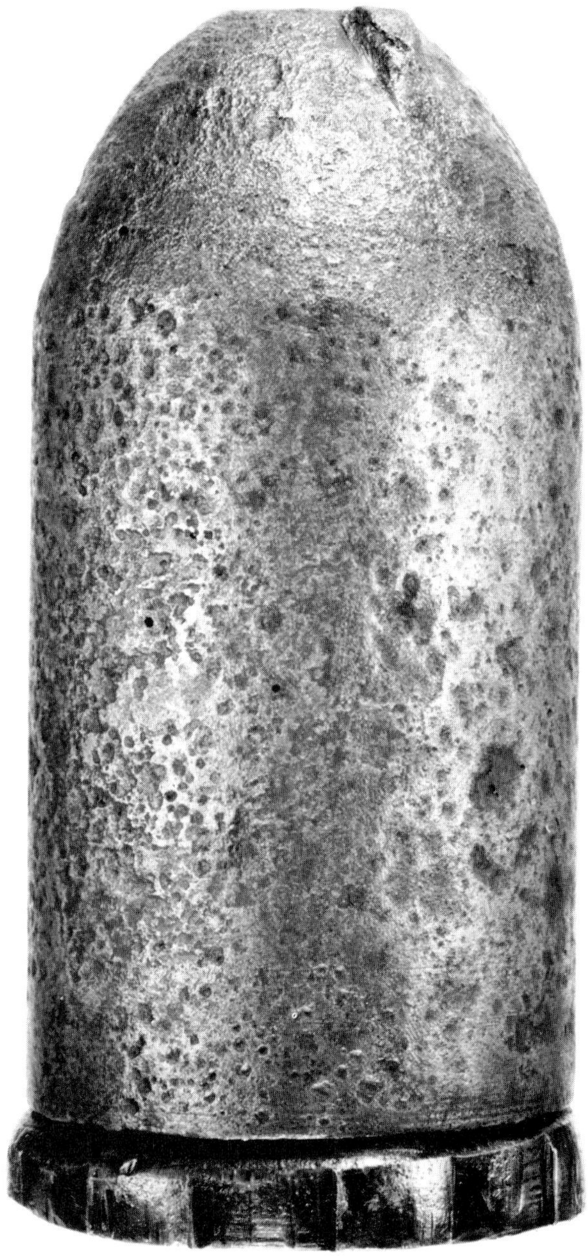

PATTERN:	Read
DIAMETER:	3.44 inches
GUN:	12-pounder Blakely rifle, 3.5-inch caliber
LENGTH:	7 1/8 inches
WEIGHT:	11 pounds, 13 ounces
TYPE:	Shell
SABOT:	Copper ring
FUSING:	Wooden fuse plug, paper time fuse

Comments:: Notice the lathe dog near the fuse hole of this projectile. The excellent seven lands and grooves of the 12-pounder (3.5-inch caliber) Blakely rifle is evident on the thick copper sabot. This Confederate-manufactured pattern replaced the limited supply of British-imported Britten projectiles. See page 101, figure III A-54, for the Britten projectile.

Fig. III A-159

Pre-war Federal

PATTERN:	Read
DIAMETER:	3.64 inches
GUN:	3.67-inch caliber rifle
LENGTH:	9 1/8 inches
WEIGHT:	10 pounds, 10 ounces
TYPE:	Blind shell
SABOT:	Soft metal sleeve
FUSING:	None, solid iron pin in nose

Comments: This projectile was patented by John B. Read on November 24, 1857, patent #18,707. It was found at the site of the test range at Fort Monroe, Virginia. On the flat base are six small holes around the outside diameter extending underneath the sabot, which aided the propellant gases in expanding the sabot to better fit the cannon's rifling.

Fig. III A-160

Confederate

PATTERN:	Read
DIAMETER:	3.61 inches
GUN:	20-pounder Parrott rifle, 3.67-inch caliber
LENGTH:	8 3/8 inches
WEIGHT:	17 pounds, 10 ounces
TYPE:	Solid shot
SABOT:	Copper ring
FUSING:	None

Comments: The Read pattern pictured here is rare in this caliber and pounder size. Note the high copper sabot and that the upper bourrelet is more distinct than the lower. Specimens have only been recovered from middle Tennessee. This projectile was fired by Morton's Battery of Forrest's Artillery.

Fig. III A-161

Confederate

PATTERN: Read
DIAMETER: 3.62 inches
GUN: 20-pounder Parrott rifle, 3.67-inch caliber
LENGTH: 10 inches
WEIGHT: 20 pounds, 8 ounces
TYPE: Case shot
SABOT: Copper ring
FUSING: Threaded fuse plug missing

Comments: The above pictured projectile was found at Gettysburg, Pennsylvania. When emptied, the case-shot material was found to consist of lead castings and bullets. It has a thick copper ring and a lathe dimple in the base. It is similar in appearance to the Federal 20-pounder Parrott projectile on page 171, figure III A-124.

Fig. III A-162

Confederate

PATTERN: Read
DIAMETER: 3.75 inches
GUN: 14-pounder James rifle, 3.8-inch caliber
LENGTH: 7 3/8 inches
WEIGHT: 12 pounds, 2 ounces
TYPE: Shell
SABOT: Copper ring
FUSING: Confederate percussion, copper anvil cap

Comments: The Confederates used captured James rifled cannon and manufactured their own ammunition for them. This specimen is an example of a rare Read used in that gun. The 3.8-inch caliber cannon was manufactured by the Confederates but in very limited quantities. Notice the excellent rifling of fifteen lands and grooves of the 14-pounder (3.8-inch caliber) James rifle; note also the West Point-style percussion fuse.

Fig. III A-163

Confederate

PATTERN: Read
DIAMETER: 4.59 inches
GUN: 4.62-inch rifle
LENGTH: 7 5/8 inches
WEIGHT: 17 pounds, 4 ounces
TYPE: Shell
SABOT: Copper ring
FUSING: Percussion (missing)

Comments: This is an extremely rare Read pattern that was used in the 4.62-inch caliber rifle. This unfired specimen was manufactured at the Selma Arsenal in Selma, Alabama.

Fig. III A-164

Federal

PATTERN: Sawyer
DIAMETER: 3.56 inches
GUN: 6-pounder Sawyer rifle, 3.67-inch caliber
LENGTH: 7 1/8 inches
WEIGHT: 15 pounds, 10 ounces
TYPE: Solid shot
SABOT: Lead jacket
FUSING: None

Comments: Sylvanus Sawyer of Fitchburg, Massachusetts, patented this style of projectile on November 13, 1855, patent #13,799. The patent date "Nov. 13, 1855" is stamped on the lead base, which is slightly concave. Note the nice rifling on the lead sabot from the 3.67-inch Sawyer rifled cannon. This specimen was found at Port Hudson, as have been most examples of this type.

Fig. III A-165

Federal

PATTERN: Sawyer
DIAMETER: 3.59 inches
GUN: 6-pounder Sawyer rifle, 3.67 inch caliber
LENGTH: 9 1/2 inches with fuse
WEIGHT: 15 pounds, 3 ounces
TYPE: Shell
SABOT: Lead jacket
FUSING: Sawyer combination

Comments: Addison M. Sawyer patented the combination fuse seen in this specimen on May 26, 1863, patent #38,699. The fuse was designed to be a time and percussion fuse. If the time fuse did not ignite from the propellant charge, then the impact was supposed to detonate the projectile. This style of fuse is also referred to as a Sawyer candlestick fuse. Stamped on the nose is "14 1/2," which is the amount of bursting charge in ounces that was to be installed at the arsenal. Several 3-inch Sawyer projectiles have been found in Virginia.

Fig. III A-166

Federal

PATTERN: Sawyer
DIAMETER: 3.61 inches
(3.83 across flanges)
GUN: 6-pounder Sawyer rifle, 3.67-inch caliber
LENGTH: 7 5/16 inches
WEIGHT: 12 pounds, 4 ounces
TYPE: Shell
SABOT: Iron flanges covered with lead jacket
FUSING: Sawyer percussion

Comments: Sylvanus Sawyer patented this projectile and fuse on November 13, 1855, patent #13,799. Sawyer shells had poor battlefield performance because the lead jacket expanded when the shell burst, thereby absorbing the shock and not allowing the iron to fragment properly. This pattern has six iron flanges under the lead covering that correspond with the rifling of the bore. The patent date is stamped on the flat base. Stamped on the nose near the fuse is "11 1/2." This information was needed so that the factory worker would know how many ounces—11 1/2—of bursting charge to load inside the projectile. Sawyer projectiles are scarce. Most of the specimens recovered have been found at Port Hudson, Louisiana, and were fired by Holcomb's Vermont Battery.

Fig. III A-167

Confederate

PATTERN:	Schenkl (copy)
DIAMETER:	2.20 inches
GUN:	2.25-inch Confederate mountain rifle
LENGTH:	8 1/16 inches
WEIGHT:	3 pounds, 12 ounces
TYPE:	Shell
SABOT:	Copper ring (missing)
FUSING:	Confederate copper fuse plug, paper time fuse

Comments: Only two specimens are known to the authors. This projectile was an attempt by the Confederates to copy the Federal Schenkl projectile. The copper fuse adapter is smaller than that of the typical Confederate fuse plug. There was a thick copper ring that served as a sabot. The other specimen, which has the sabot intact, is located in the West Point Museum.

Fig. III A-168

Federal

PATTERN: Schenkl
DIAMETER: 2.92 inches
GUN: 3-inch wrought iron (ordnance) rifle
LENGTH: 9 3/16 inches
WEIGHT: 8 pounds
TYPE: Shell
SABOT: Papier-mâché (missing)
FUSING: Schenkl percussion, brass

Comments: This is the most common Schenkl pattern. There are seven vertical raised ribs on the tapered cone. Cyrus Alger & Company of Boston, Massachusetts, manufactured more than 100,000 of these 3-inch Schenkls, which were purchased by the Federal ordnance department during the war. Stamped on the brass percussion fuse is "J.P. SCHENKL PAT. OCT. 16, 1861." The actual patent date was October 15, 1861, patent #33,495. According to Abbot's *Siege Artillery in the Campaigns Against Richmond,* the Schenkl percussion fuse was 82% effective.

Fig. III A-169

Federal

PATTERN:	Schenkl
DIAMETER:	2.90 inches
GUN:	3-inch wrought iron (ordnance) rifle
LENGTH:	10 1/8 inches
WEIGHT:	8 pounds without sabot
TYPE:	Case shot
SABOT:	Papier-mâché
FUSING:	Schenkl combination

Comments: The rounded nose on this pattern indicates it is loaded with lead case-shot balls. Because the papier-mâché sabot fit so tightly in the bore, no flame could pass to light the conventional time fuse. A self-igniting fuse was used in an attempt to correct this problem. However, the combination fuse worked very poorly. During the campaigns against Richmond, Brevet Brigadier General Henry L. Abbot wrote that the Schenkl combination fuse was so inferior that only 55% exploded the projectile. This shell has seven raised ribs on the tapered cone. Please note the original papier-mâché sabot on this non-excavated example. The lower zinc band, which helped decrease humidity-related swelling, is missing. See page 42 for an explanation of the Schenkl combination fuse.

Fig. III A-170

Federal converted to Confederate

PATTERN: Schenkl
DIAMETER: 2.93 inches
GUN: 3-inch rifle
LENGTH: 9 inches
WEIGHT: 8 pounds, 2 ounces
TYPE: Shell
SABOT: Wooden cone (missing)
FUSING: Confederate copper fuse plug, paper time fuse

Comments: This Federal Schenkl was captured by the Confederates and sent to an arsenal where it was equipped with a wooden sabot to replace the papier-mâché sabot. The Confederates did not have the means to manufacture the papier-mâché sabot. The iron pin at the base secured the wooden sabot. The Schenkl brass percussion fuse was replaced with a Confederate copper fuse plug for a paper time fuse.

Fig. III A-171

Federal

PATTERN: Schenkl
DIAMETER: 3.55 inches
GUN: 3.67-inch caliber rifle
LENGTH: 8 13/16 inches
WEIGHT: 10 pounds, 5 ounces
TYPE: Shell
SABOT: Papier-mâché (missing)
FUSING: Schenkl percussion, brass

Comments: This specimen has six narrow ribs on the tapered cone. The brass Schenkl percussion fuse was patented by John P. Schenkl on October 15, 1861, patent #33,495. However, stamped on this specimen's fuse is "J.P. SCHENKL PAT. OCT. 16, 1861." An improved fuse model was patented by J.P. Schenkl on August 19, 1862, patent #36,236.

Fig. III A-172

Federal

PATTERN: Schenkl
DIAMETER: 3.62 inches
GUN: 3.67-inch caliber rifle
LENGTH: 8 11/16 inches
WEIGHT: 10 pounds, 11 ounces
TYPE: Shell
SABOT: Papier-mâché (missing)
FUSING: Schenkl percussion, brass

Comments: This specimen has six recessed areas that are wider than usual. See page 218, figure III A-171, for fuse comments.

Fig. III A-173

Confederate

PATTERN:	Schenkl (copy)
DIAMETER:	3.63 inches
GUN:	3.67-inch caliber rifle
LENGTH:	8 1/2 inches
WEIGHT:	13 pounds, 7 ounces
TYPE:	Solid shot
SABOT:	Wooden cone
FUSING:	None

Comments: This specimen is of Trans-Mississippi manufacture and was found at a site where a caisson tumbled off a bridge at Provencal, Louisiana, during Banks' Red River Campaign. There are twelve thin raised ribs on the tapered cone. The wooden sabot was held on by a thin iron plate, which was affixed by placing it over a small pin cast into the base of the shell and hammering the pin flat. The Confederates lacked the formula for the manufacture of papier-mâché and used wood in its place.

Fig. III A-174

Confederate

PATTERN: Schenkl (copy)
DIAMETER: 3.60 inches
GUN: 3.67-inch caliber rifle
LENGTH: 8 11/16 inches
WEIGHT: 13 pounds
TYPE: Solid shot
SABOT: Wooden cone (missing)
FUSING: None

Comments: This Schenkl pattern has twelve ribs on the tapered cone and is missing the wooden sabot. It was recovered from same site as the specimen on page 220, figure III A-173, and the same comments apply. There was no pin protruding from the base, as seen on previous specimen.

Fig. III A-175

Confederate

PATTERN: Schenkl (copy)
DIAMETER: 3.59 inches
GUN: 3.67-inch caliber rifle
LENGTH: 8 1/4 inches
WEIGHT: 9 pounds, 10 ounces
TYPE: Shell
SABOT: Wooden cone (missing)
FUSING: Wooden fuse plug, paper time fuse

Comments: This is the shell version of the specimen noted on page 221, figure III A-174, and the same comments apply.

Fig. III A-176

Federal

PATTERN:	Schenkl
DIAMETER:	3.60 inches
GUN:	20-pounder Parrott rifle, 3.67-inch caliber
LENGTH:	11 3/16 inches
WEIGHT:	15 pounds, 6 ounces
TYPE:	Shell
SABOT:	Papier-mâché (missing)
FUSING:	Schenkl percussion, brass

Comments: The six ribs on this specimen are of an unusual ratcheted style. This pattern typically has six raised ribs similar to the specimen on page 218, figure III A-171. Although seldom seen, 20-pounder Schenkl projectiles have been recovered in Virginia.

Fig. III A-177

Confederate

PATTERN: Selma
DIAMETER: 2.43 inches
GUN: 2.5-inch Blakely rifle
LENGTH: 7 1/4 inches
WEIGHT: 5 pounds, 14 ounces
TYPE: Shell
SABOT: Copper plate
FUSING: Wooden fuse plug, paper time fuse

Comments: Projectiles of this pattern are named after their attributed site of manufacture: Selma Arsenal in Selma, Alabama. In the groove between the high copper sabot and the cylindrical portion of the projectile body can be seen remnants of a hemp-like material. This material was originally greased, thus serving as a lubricant for the cannon bore.

Fig. III A-178

Confederate

PATTERN: Selma
DIAMETER: 2.95 inches
GUN: 3-inch rifle
LENGTH: 7 5/8 inches
WEIGHT: 8 pounds, 8 ounces
TYPE: Shell
SABOT: Copper plate
FUSING: Wooden fuse plug, paper time fuse

Comments: Note the wide lubrication groove between the high copper sabot and the distinct bottom bourrelet. Remnants of the lathe dog can be seen on the projectile's nose. A lathe dimple is found in the center of the copper base. This pattern of projectile is rare.

Fig. III A-179

Confederate

PATTERN: Selma
DIAMETER: 2.96 inches
GUN: 3-inch rifle
LENGTH: 7 1/2 inches
WEIGHT: 7 pounds, 8 ounces
TYPE: Shell
SABOT: Copper plate
FUSING: Wooden fuse plug, paper time fuse

Comments: This specimen is similar in appearance to the Read pattern on page 195, figure III A-148. The copper sabot is continuous across the bottom of the recessed base. The bourrelets are narrower and the lubrication groove is much smaller than the projectile on page 225, figure III A-178. This projectile was manufactured by the Selma Arsenal in Alabama.

Fig. III A-180

Confederate

PATTERN:	Selma
DIAMETER:	3.21 inches
GUN:	3.3-inch Confederate rifle
LENGTH:	6 3/4 inches
WEIGHT:	10 pounds, 11 ounces
TYPE:	Solid shot
SABOT:	Copper plate
FUSING:	None

Comments: Stamped on the nose and on the bottom of the copper sabot is the letter "G," which indicates Selma Arsenal manufacture. This is a typical example of this pattern. Similar specimens have the letter "R" stamped on the nose and sabot. Numbers are often found stamped into the upper bourrelet. Projectiles of this variety have been recovered in Milledgeville, Georgia. A rounded-nose variety also exists. The base of the solid-copper sabot is concave. See also the comments on page 201, figure III A-154.

Fig. III A-181

Confederate

PATTERN:	Selma
DIAMETER:	3.60 inches
GUN:	3.67-inch caliber rifle
LENGTH:	5 11/16 inches
WEIGHT:	10 pounds, 6 ounces
TYPE:	Solid shot
SABOT:	Copper plate, cloverleaf
FUSING:	None

Comments: Notice the heavy lathe marks on the bourrelets as well as the copper sabot. Most often, fired projectiles of this pattern will have the sabot distorted or partly missing. This pattern tended to tumble due to its incorrect length-to-diameter ratio for the caliber of the gun. A good ratio is between 2 1/2 to 3 times the diameter or caliber for the length of the projectile.

Fig. III A-182

Confederate

PATTERN: Selma
DIAMETER: 3.62 inches
GUN: 20-pounder Parrott rifle, 3.67-inch caliber
LENGTH: 8 15/16 inches
WEIGHT: 16 pounds, 6 ounces
TYPE: Segmented shell
SABOT: Copper plate
FUSING: Wooden fuse plug, paper time fuse

Comments: This projectile has a segmented interior in the form of a six-sided prism. Upon successful detonation, it was designed—ideally—to break the shell into equal-size fragments except for the nose and the base. This specimen is unfired.

Fig. III A-183

Great Britain

PATTERN: Whitworth
DIAMETER: 1.68 inches across flats,
1.82 inches outer diameter
GUN: 3-pounder Whitworth,
1.7-inch caliber
LENGTH: 5 9/16 inches
WEIGHT: 2 pounds, 11 ounces
TYPE: Shell
SABOT: None
FUSING: None, brass shipping plug

Comments: Joseph Whitworth of Manchester, England, patented this pattern of projectile dated April 23, 1855, Great Britain patent #903. His patent states, " . . . projectiles [are] made in a spirally shaped form, so as to fit exactly the interior of a piece of ordnance or fire-arm made of a corresponding form, thereby enabling harder metals to be used, economising the force hitherto expended in pressing the projectile into rifle grooves, and diminishing the loss by windage and friction." No battlefield recoveries of the pictured specimen are known to the authors. The use of this projectile during the American Civil War is unlikely. Note the non-tapered base, which was felt to contribute to an unstable flight. As a result, the base in other Whitworth patterns was tapered to correct this defect.

Fig. III A-184

Great Britain

PATTERN: Whitworth
DIAMETER: 2.15 inches across flats
GUN: 6-poundcr Whitworth riflc, 2.17-inch caliber
LENGTH: 7 1/4 inches
WEIGHT: 5 pounds, 15 ounces
TYPE: Solid shot
SABOT: None
FUSING: None

Comments: Joseph Whitworth patented an improvement on his Whitworth projectile dated December 30, 1859, Great Britain patent #2990. His patent states, "The rear part is made to taper, and the weight, form, and degree of convergence given to it must be determined by and adapted to those of the fore part." This improvement gave it more favorable accuracy, velocity, and range. This caliber was used to a very limited extent. Most Whitworth projectiles used were bolts. This was due to the small ineffective bursting cavity in the shells, which rendered them less effective. These are commonly referred to as 6-pounder Whitworth projectiles.

Fig. III A-185

Great Britain

PATTERN:	Whitworth
DIAMETER:	2.15 inches across flats
GUN:	6-pounder Whitworth rifle, 2.17-inch caliber
LENGTH:	9 inches
WEIGHT:	5 pounds, 13 ounces
TYPE:	Shell
SABOT:	None
FUSING:	Threaded fuse missing

Comments: The 6-pounder 2.17-inch caliber Whitworth rifle saw very little battlefield use. Shells of this pattern are extremely rare. The small bursting cavity made the projectile fragment poorly. Whitworth projectiles were not only accurate, but could achieve distances that no other Civil War field artillery projectiles could.

Fig. III A-186

Confederate

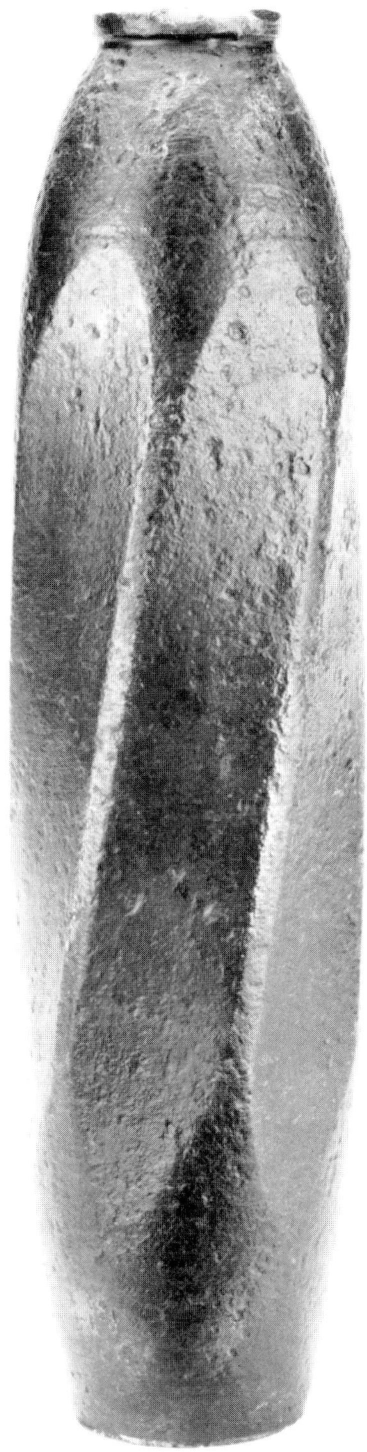

PATTERN: Whitworth
DIAMETER: 2.14 inches across flats
GUN: 6-pounder Whitworth rifle, 2.17-inch caliber
LENGTH: 9 1/16 inches
WEIGHT: 7 pounds, 6 ounces
TYPE: Shell
SABOT: None
FUSING: Confederate copper fuse plug, paper time fuse

Comments: This is an attempt by the Confederates to copy the difficult-to-obtain British-imported Whitworth ammunition. Note the Confederate copper fuse plug and the crude machining marks around the nose of this projectile.

Fig. III A-187

Great Britain

PATTERN:	Whitworth
DIAMETER:	2.73 inches across flats
GUN:	12-pounder Whitworth rifle, 2.75-inch caliber
LENGTH:	9 7/8 inches
WEIGHT:	12 pounds, 11 ounces
TYPE:	Solid shot
SABOT:	None
FUSING:	None

Comments: Joseph Whitworth patented this projectile dated April 23, 1855, Great Britain patent #903. His patent information stated: ". . . to projectiles made in a spirally shaped form, so as to fit exactly the interior of a piece of ordnance or fire-arm made of a corresponding form." He later patented an improvement in projectiles dated December 30, 1859, Great Britain patent #2990, which stated: ". . . in so forming the projectile that the rear part may have a proper relation to the fore part with regard to their respective shapes and weights. The shape of the fore part depends upon the purpose for which the projectile is to be employed, and may be more or less pointed or curved, or flat fronted. The rear part is made to taper, and the weight, form, and degree of convergence given to it must be determined by and adapted to those of the fore part. The parts thus mutually adapted present their surfaces to the action of the air under conditions most favourable to the accuracy, velocity, and range of the projectile."

This is the most common Whitworth bolt projectile manufactured. Both Federal and Confederate forces used the 12-pounder Whitworth, although the majority were fired by the Confederates. This is the most common Whitworth bolt pattern manufactured. Notice the machining marks on the projectile body between the flat surfaces.

Fig. III A-188

Great Britain

PATTERN:	Whitworth
DIAMETER:	2.71 inches across flats
GUN:	12-pounder Whitworth rifle, 2.75-inch caliber
LENGTH:	9 inches
WEIGHT:	8 pounds, 8 ounces
TYPE:	Segmented shell
SABOT:	None
FUSING:	Threaded fuse missing

Comments: This Whitworth was loaded through the removable nose, which allowed iron segments to be arranged around the powder charge. Three pins secured the nose. No battlefield recoveries in this pattern and size are known to the authors.

Fig. III A-189

Great Britain

PATTERN:	Whitworth
DIAMETER:	2.73 inches across flats
GUN:	12-pounder Whitworth rifle, 2.75-inch caliber
LENGTH:	10 7/8 inches
WEIGHT:	12 pounds
TYPE:	Shell
SABOT:	None
FUSING:	Confederate copper fuse plug, paper time fuse

Comments: Note the excellent British workmanship. Although of British origin, this specimen is fitted with a Confederate copper time fuse plug.

III. A Pictorial Study
B. Non-Combat / Post-War

The projectiles in this section are considered to be non-combat-proven Civil War era or of post-war manufacture. Any corrections to this section with proof of battlefield use would be gratefully appreciated by all collectors in the field.

The post-war section is representative and is not meant to be inclusive of all post-war ordnance available. The purpose of this section is to make the student of Civil War artillery aware of the projectiles that could be potentially confused with Civil War-era battlefield-proven ordnance.

Fig. III B-1

Federal

PATTERN: Absterdam
DIAMETER: 2.94 inches
GUN: 3-inch rifle
LENGTH: 6 7/8 inches
WEIGHT: 9 pounds, 5 ounces
TYPE: Solid shot
SABOT: Brass ring
FUSING: None

Comments: John Absterdam's patent for projectiles was dated February 23, 1864, patent #41,668. Stamped on the brass sabot is "ABSTERDAM'S PATENT FEB. 23, 1864." The letters "JPK," stamped on the side of the projectile, are thought to be inspector's initials. The sabot has six cuts to aid in expansion and extends 1/4" below the base of the projectile. There are no known battlefield recoveries.

Fig. III B-2

Federal

PATTERN: Absterdam
DIAMETER: 2.94 inches
GUN: 3-inch rifle
LENGTH: 7 1/2 inches
WEIGHT: 9 pounds, 4 ounces
TYPE: Shell
SABOT: Lead cup
FUSING: Absterdam (missing)

Comments: This pattern has only one bearing surface. Stamped on the bottom of the lead cup is "PATENTED S & CO AUGUST 12, 1862." John Absterdam of New York, New York, patented an improvement in banding and covering projectiles on November 7, 1865, patent #50,783. There are no known battlefield recoveries of this projectile in 3-inch caliber.

Fig. III B-3

Federal

PATTERN: Absterdam
DIAMETER: 2.94 inches
GUN: 3-inch rifle
LENGTH: 8 1/2 inches
WEIGHT: 9 pounds, 13 ounces
TYPE: Shell
SABOT: Brass ring
FUSING: Absterdam brass fuse plug, paper time fuse

Comments: This is the most common Absterdam pattern. Most of these projectiles are seen without fuses. Some of the shells have been noted with case shot. There are no known battlefield recoveries.

Fig. III B-4

Federal

PATTERN: Boekel
DIAMETER: 2.95 inches
GUN: 3-inch ordnance
LENGTH: 7 3/4 inches
WEIGHT: 8 pounds, 10 ounces
TYPE: Shell
SABOT: Lead jacket
FUSING: Threaded fuse missing

Comments: William Boekel of Pennsylvania first patented this style of projectile on September 16, 1862, patent #36,449. Boekel also patented on April 12, 1864, patent #42,268, an improved projectile, as shown in this specimen. The lead jacket is covered by a thin brass sleeve. This sleeve was to help prevent the lead from fouling the bore of the rifle. The lead jacket is concave on the base. The Boekel projectile was tested at the Washington Arsenal test range in October of 1864, with good remarks. There are no known battlefield recoveries of this projectile.

Fig. III B-5

Federal

PATTERN: Butler
DIAMETER: 3.15 inches
GUN: Unknown
LENGTH: 9 inches
WEIGHT: 12 pounds
TYPE: Solid shot
SABOT: Brass ring, Butler's patent
FUSING: None

Comments: John G. Butler patented this projectile on September 26, 1871, patent #119,313. The brass sabot was screwed onto the iron base of the projectile in an attempt to prevent the sabot from turning or slipping. There is a space between the sabot and the projectile body to receive the expansion gases, which allowed the sabot to enter the rifling.

Fig. III B-6

Federal

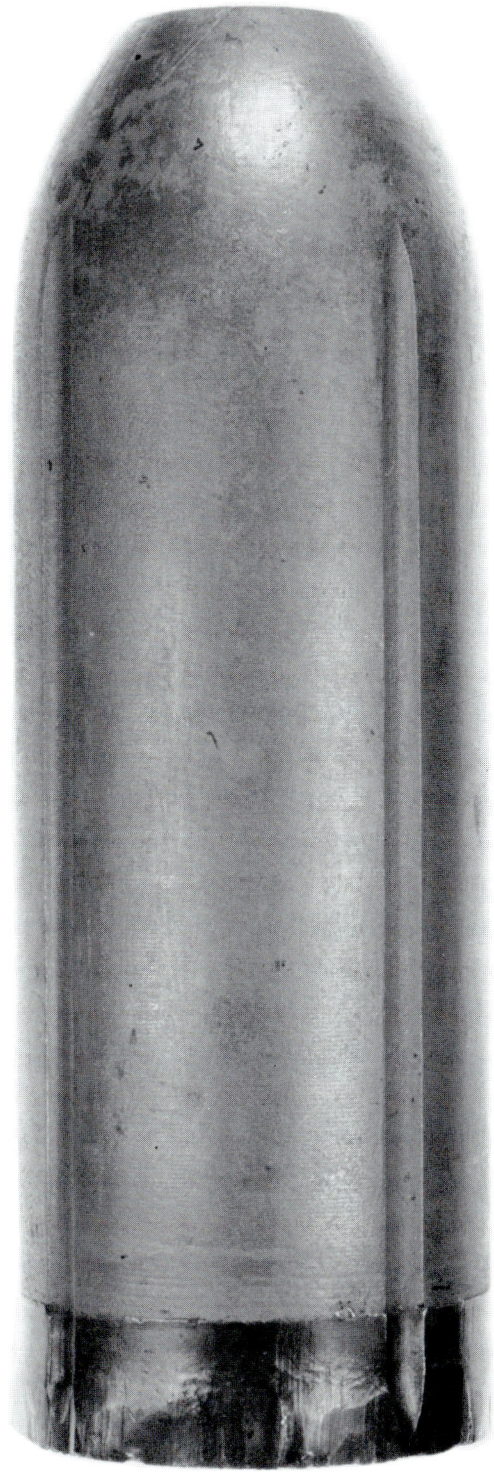

PATTERN:	Butler
DIAMETER:	2.96 inches
GUN:	3-inch wrought iron (ordnance) rifle
LENGTH:	8 1/2 inches
WEIGHT:	9 pounds, 14 ounces
TYPE:	Shell
SABOT:	Brass ring, Butler's patent
FUSING:	Threaded time fuse plug (missing)

Comments: This projectile appears to be a variant of the Federal Absterdam. However, patent information suggests that the sabot is most likely covered under the Butler patent noted opposite. Whatever the precise origin, this projectile was never combat proven. Note the four flame grooves that run the entire length of the projectile body. The brass sabot has the impressions of the seven lands and grooves of the Federal 3-inch wrought iron (ordnance) rifle.

Fig. III B-7

State Of New York

PATTERN:	Delafield
DIAMETER:	3.61 inches
GUN:	20-pounder Delafield rifle, 3.67-inch caliber
LENGTH:	9 3/16 inches
WEIGHT:	13 pounds, 2 ounces
TYPE:	Shell
SABOT:	Malleable iron
FUSING:	Parrott zinc fuse plug, paper time fuse

Comments: This projectile, along with a corresponding rifled cannon, was invented by Lieutenant Colonel Richard Delafield for the State of New York. A schematic drawing Delafield prepared of his cannon is dated September 11, 1861. Thirteen of the Delafield cannon were delivered to the Commissary General of the State of New York on March 18, 1862. The base of this projectile has five cast raised flanges that correspond to the Delafield cannon's rifling. This specimen is similar in appearance to the Federal 20-pounder Parrott projectile. There have been no known combat finds, and battlefield use is unlikely. See page 245, figure III B-8, for a comparison of the base to a Federal 20-pounder Parrott projectile.

Fig. III B-8

Delafield and Parrott

Comparison views

Comments: The 20-pounder (3.67-inch caliber) Delafield projectile is on the reader's left and the 20-pounder (3.67-inch caliber) Parrott projectile is on the right. Note the five distinct flanges on the base of the Delafield, with a deep recessed base. The Parrott has five pre-stamped projections in the wrought iron sabot that correspond to the five grooves of the 20-pounder Parrott rifle. These projectiles are often confused with each other. See page 244, figure III B-7, for a Delafield specimen and page 171, figure III A-124, for a Parrott example.

Fig. III B-9

Federal

PATTERN: Dyer
DIAMETER: 2.47 inches
GUN: Unknown
LENGTH: 7 1/4 inches
WEIGHT: 8 pounds, 2 ounces
TYPE: Solid shot
SABOT: Lead cup
FUSING: None

Comments: This projectile is of post-war manufacture. A longer solid-shot variety is 9 1/4 inches long.

Fig. III B-10

Federal

PATTERN: Dyer
DIAMETER: 2.96 inches
GUN: 3-inch rifle
LENGTH: 8 3/16 inches
WEIGHT: 7 pounds, 3 ounces
TYPE: Shell
SABOT: Lead cup
FUSING: Taylor fuse plug, paper time fuse

Comments: The fuse in the pictured specimen was an improvement over the standard time fuse. The addition of holes around the fuse opening allowed the ignition shield to direct the flame to the paper time fuse. The fuse is stamped "TAYLOR" and bears the patent date "APRIL 11, 1865." There are no known battlefield recoveries of this pattern.

Fig. III B-11

Federal

PATTERN: Emery
DIAMETER: 3.62 inches
GUN: 20-pounder Parrott rifle, 3.67-inch caliber
LENGTH: 8 1/2 inches including bolt
WEIGHT: 15 pounds, 10 ounces
TYPE: Solid shot
SABOT: Brass ring (missing)
FUSING: None

Comments: A.H. Emery patented this unusual projectile on March 17, 1863, patent #37,906. He patented a fuse for the shell version of his projectile on December 8, 1863, patent #40,828. It had a brass ring sabot that was attached to the projectile after casting. This pattern is extremely rare, and no battlefield finds have been noted to date.

Fig. III B-12

Federal

PATTERN: Eureka
DIAMETER: 2.96 inches
GUN: 3-inch rifle
LENGTH: 8 1/4 inches
WEIGHT: 8 pounds, 9 ounces
TYPE: Shell
SABOT: Brass ring
FUSING: Eureka brass fuse plug, paper time fuse

Comments: The Eureka projectile was manufactured by the American Projectile Company in New York and was test fired by the Washington Arsenal as early as May 17, 1864. The brass fuse has a screwdriver slot and spanner holes. Stamped near the fuse are the letters "JH." The brass sabot consists of a concave disk and a short cylindrical cone attached to the base of the projectile with a combination of rivets.

Fig. III B-13

Federal

PATTERN:	Hotchkiss
DIAMETER:	2.94 inches
GUN:	3-inch wrought iron (ordnance) rifle
LENGTH:	6 5/8 inches
WEIGHT:	9 pounds, 1 ounce
TYPE:	Shell
SABOT:	Lead band
FUSING:	Hotchkiss combination time and percussion fuse

Comments: Benjamin Hotchkiss patented this improvement for the Hotchkiss projectile on May 16, 1865, patent #47,725. A wooden disk was inserted between the base cup and the base of the nose to decrease the risk of premature fracture of the shell during initial firing. The insufficient burning time of the standard-style 5 1/4-second Bormann time fuse led to the development of the 14-second version for rifled ordnance, which was patented on December 24, 1867, patent #72,494. The letters "TB" are stamped into the lead band. See page 23, figure II B-9, for a cross-sectional view.

Fig. III B-14

Federal

PATTERN: Hotchkiss
DIAMETER: 2.94 inches
GUN: 3-inch wrought iron (ordnance) rifle
LENGTH: 6 5/16 inches
WEIGHT: 8 pounds, 14 ounces
TYPE: Shell
SABOT: Lead band
FUSING: Hotchkiss combination time and percussion fuse

Comments: The base cup on this pictured specimen is flat and is of post-war manufacture. The protective cover of the fuse has been removed to show the flammable material that was used to help direct flame to the punched Bormann fuse. The same comments apply to this projectile as that on page 250, figure III B-13.

Fig. III B-15

Federal

PATTERN: Parrott
DIAMETER: 2.85 inches
GUN: 10-pounder Parrott rifle, 2.9-inch caliber
LENGTH: 6 1/2 inches
WEIGHT: 10 pounds, 4 ounces
TYPE: Solid shot
SABOT: Lead cup
FUSING: None

Comments: This is thought to be an early experimental Parrott projectile. The lead sabot has three 1.42 "-wide flanges that correspond to the 10-pounder Parrott (2.9-inch caliber) rifling. The sabot is 1 1/2" tall and concave on the base. The numbers "8," "10," and "12" are stamped into the iron body. Note the fine machine marks located on the cylindrical iron body. It is extremely rare for Parrott 10-pounder (2.9-inch caliber) or 3-inch caliber projectiles to have lead sabots. There are no known battlefield recoveries.

Fig. III B-16

Federal

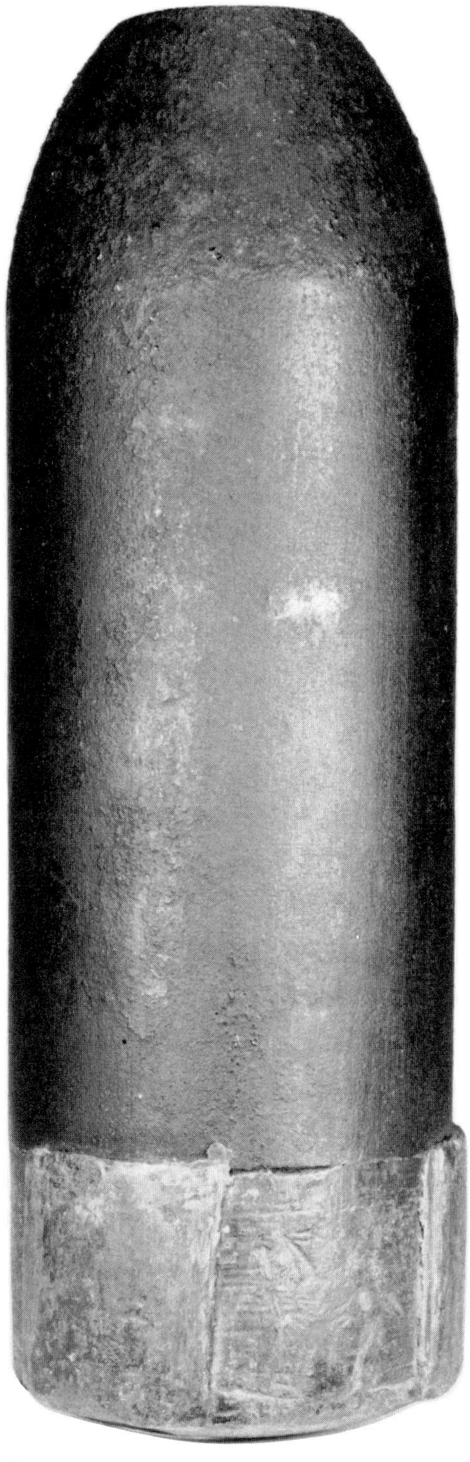

PATTERN: Parrott
DIAMETER: 2.86 inches
GUN: 10-pounder Parrott rifle, 2.9-inch caliber
LENGTH: 8 5/8 inches
WEIGHT: 9 pounds, 14 ounces
TYPE: Shell
SABOT: Lead cup
FUSING: Threaded fuse (missing)

Comments: This is thought to be an early experimental Parrott projectile. The lead sabot differs from that of the specimen on page 252, figure III B-15. The pattern pictured above has a rounded base compared to the concave base mentioned. The concave base probably expanded into the rifling too much and the rounded base was developed from this trial. The 1 7/8"-tall lead sabot has three 1.42"-wide flanges that correspond to the 10-pounder (2.9-inch caliber) Parrott rifling. A Parrott zinc time fuse plug screws easily into the fuse threads.

Fig. III B-17

Federal

PATTERN: Stafford
DIAMETER: 2.96 inches
GUN: 3-inch rifle
LENGTH: 7 7/8 inches
WEIGHT: 10 pounds, 11 ounces
TYPE: Solid shot
SABOT: Brass disk
FUSING: None

Comments: Charles W. Stafford of New York, patented this projectile on December 20, 1864, patent #45,567. The brass disk sabot has three corresponding flanges that fit into three grooves on the tapered iron base of the projectile. The brass disk is affixed to the base of the projectile with a six-sided bolt.

Fig. III B-18

Federal

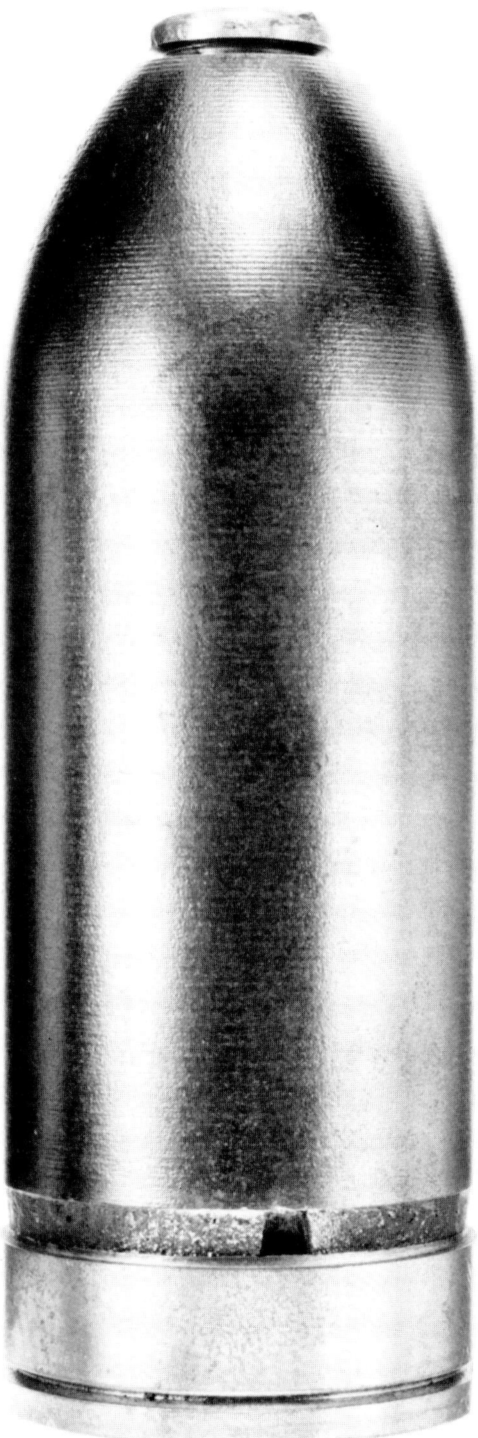

PATTERN: Stafford
DIAMETER: 2.96 inches
GUN: 3-inch rifle
LENGTH: 8 1/4 inches
WEIGHT: 8 pounds, 7 ounces
TYPE: Shell
SABOT: Brass disk
FUSING: Hotchkiss brass fuse plug, paper time fuse

Comments: This is the shell variety of the patented C.W. Stafford projectile. The same comments apply as those on the specimen on page 254, figure III B-17, opposite. There are no known battlefield recoveries of the Stafford pattern.

Fig. III B-19

Post-War Souvenir

PATTERN: Whitworth
DIAMETER: 2.79 inches across flats
GUN: None (see comments below)
LENGTH: 11 1/8 inches to top of shell
WEIGHT: 10 pounds, 7 ounces
TYPE: Shell
SABOT: None
FUSING: Schenkl percussion, brass

Comments: This projectile was thought to have been manufactured as a souvenir for relic collectors sometime after the Civil War ended. All specimens found to date are too large in diameter for any known Whitworth cannon. Note the crude casting flaws and mould seam that runs the entire length of the projectile. Examples of Whitworth bolts were known to have been made at monuments at Gettysburg National Military Park. The park specimens are larger in diameter than those of the Civil War period.

Fig. III B-20

French

PATTERN: Unknown
DIAMETER: 3.30 inches of iron body
GUN: Unknown
LENGTH: 6 1/2 inches
WEIGHT: 9 pounds
TYPE: Shell
SABOT: Lead studs
FUSING: Threaded fuse (missing)

Comments: According to a letter written by Lieutenant Colonel Richard Delafield on May 29, 1960, this is a French rifled projectile. A similar specimen was picked up on the battlefield of Solforina, Italy, in 1860. Based on the diameter of the projectile and not the studs, no Confederate or Federal cannon known to the authors could have fired this projectile during the American Civil War. This projectile has been confused with the known Armstrong patterns, which had three rows of copper studs. There are two rows of six lead studs, for a total of twelve studs.

V. Valuation Criteria
What Is It Worth?

I am often asked, "Why is this shell more expensive than that one?" To help answer this important question, I have listed below the factors that should be used to determine the relative value of a Civil War projectile.

Condition

Condition is extremely important because an uncommon projectile in poor condition can be less desirable than a common projectile in excellent condition.

The projectile's overall appearance must be considered in value determination. A projectile in *poor* condition will exhibit large, deep pitting, heavy metal flaking, and is in overall unstable condition. *Fair* projectiles will have medium pitting, not greater than 1/4" deep or wide, and will be in overall stable condition. A projectile in *good* condition—the average condition of excavated specimens—will have light to medium pitting, not exceeding an average of 1/8" deep or wide, and will have some surface flaking. *Fine* condition projectiles exhibit a nearly smooth surface, with only slight pitting not greater than 1/16" deep or wide. Projectiles in *excellent* condition—usually non-excavated specimens or early battlefield pickups—will be in a pristine state.

Whether the projectile is cleaned, disarmed (if it is a shell), professionally restored, and coated to retard rust are also factors used to determine value.

Rarity

A projectile that was manufactured in large quantities and is easily obtained is considered *very common*. The 12-pounder solid shot is an excellent example of this category. A projectile which is found in reasonable quantities but requires some effort to obtain is deemed *common*. The Federal 10-pounder Parrott is a common projectile. A projectile that is found less often but is still available is considered *less common*. Examples are the Federal 3.8-inch James sub-pattern II and 3-inch Confederate Read projectiles. A projectile that is hard to come by but still occasionally found is considered *rare*; for example, Federal and Confederate stands of grape or the 3-inch Confederate Brooke projectiles. A projectile which was manufactured in small quantities and is almost never available is considered *extremely rare*. Examples are the 3-inch Armstrong projectile or the Confederate 3-inch Mullane (Tennessee sabot) side-loader. A projectile that is one of a kind or one of only two known is considered *unique*. The wing shot is an example of a unique projectile.

No authors or collectors can possibly assign an accurate numeric rarity scale: the immense number of projectiles in museums, in private collections, and the unknown quantities yet to be recovered render any rarity scale inaccurate at the outset. A rarity scale will be at best subjective, the result of that particular author's or collector's experience and impressions. We have, therefore, made no attempt to classify projectiles in this work within the limiting range of a rarity scale.

Completeness

The significant questions to consider are:
(1) Was the projectile designed to have a fuse? If so, is the fuse present and intact?
(2) Was the projectile manufactured with a sabot? If so, is the sabot intact?
(3) Does the projectile body have any attached bolts, nuts, plates, nails, or screws? If so, are they broken, missing, or otherwise damaged?

Type of projectile

A solid shot (sometimes known as a bolt in rifled projectiles), shell, canister, and/or stand of grape will have different values. Some solid-shot patterns can bring a premium, whereas the same pattern with a fuse may be of more or less value, depending upon the fuse type.

Type of fuse, sabot, or interior

The type of fuse used will greatly affect the value. As a rule, projectiles with percussion, concussion, and experimental fuses will bring more than those with time fuses. A projectile with a brass or copper sabot will be more desirable than one with an iron sabot. The variety of the sabot within the patterns will also affect the projectile's value. An interior that is polygonal or segmented will bring more than a smooth-interior specimen.

Caliber or size

The most common, extensively used field artillery calibers were the following: for rifled projectiles, 2.9-inch (10-pounder) and 3-inch; for smoothbore projectiles, 4.62-inch (12-pounder). Larger, seldom-seen smoothbore projectiles will demand higher prices. Odd or limited production rifled projectiles in unusual calibers will command a higher premium. The 2.25-, 2.6-, 2.75- and 3.3-inch are examples of these projectiles.

Excavated or non-excavated

In general, a non-excavated projectile will be more desirable than the same projectile in an excavated condition. The exceptions are those fired projectiles with good rifling and in excellent condition. In addition, a late-war projectile will bring more if it is an excavated specimen. The 3-inch Hotchkiss with the 14-second Bormann fuse is one example of a late-war projectile that brings a premium if it is excavated.

Aesthetic appeal

A projectile with visual appeal will also have desirability. The way the rifling looks on the sabot, the condition of the surface of the metal, and the patina will add to or subtract from the value.

Geographic location

A specimen from a particular battlefield (Gettysburg, Petersburg, Vicksburg, etc.) will be of special interest to a person who collects artifacts from that area. Projectiles from specific campaigns, mountains, rivers, cities, and battlefields can, if documented, add considerably to the value of a projectile.

Affiliation with a unit

Unit histories, combined with a proven geographical association, will add to the value of any projectile. In some cases, Confederate projectiles that were fired by well-known artillery units will bring a premium price. Some units used only a particular size caliber or pattern of a projectile, and this adds to the specimen's desirability.

Country of origin

In many cases, the Confederates made copies of or altered Federal projectiles. As a general rule, these are more desirable than the same style of Federal projectiles. These copied projectiles bring a premium price over their Federal counterparts. Although both sides had access to British manufactured projectiles, the majority were used by the Confederates. An example of this is the British Whitworth projectile found on page 234, figure III A 187.

When considering the purchase of a projectile, all the above factors should be considered. Even though a projectile may be extremely rare, condition and desirability along with aesthetic appeal can determine the ultimate value.

It is hoped that this guide will help the collector to assess the relative value of a projectile and to make an intelligent decision regarding its purchase or sale.

—*Jack W. Melton, Jr.*

Appendix A
Caliber-to-Pounder Relationship

We have asked C. Peter Jørgensen, celebrated author and publisher of *The Artilleryman* magazine, to help explain to the reader the sometimes confusing caliber-to-pounder relationship. His remarks follow:

One of the most confusing aspects of Civil War projectile collecting involves the conversion of bore diameter to "pounder" equivalents when comparing smoothbore and rifled ammunition.

The U.S. Army didn't adopt any rifled cannon until the outbreak of the Civil War, although a number of manufacturers had successfully produced rifled muzzleloading artillery pieces. All cannon were smoothbore and were designated according to the weight of the solid shot they could fire. (Even howitzers and naval shellguns, which weren't intended for solid shot, carried the "pounder" designation based on their bore diameter.)

Here we explore one of the most confusing aspects of collecting Civil War projectiles.

Theoretically, a rifled cannon of the same bore diameter as a smoothbore would fire a projectile twice as heavy. Thus, the Model 1841 bronze 6-pounders of 3.67-inch bore were referred to as "12-pounder rifles" in many instances. They were rifled under the system invented and patented by General Charles T. James, and some of James' projectiles, including one version of his solid shot, did weigh 12 pounds.

But most smoothbores converted to rifles under the James system were rebored to 3.8-inches to make the bore interior perfectly smooth before rifling. Only a few 3.67-inch rifles exist, notably some made in 1861 directly as rifles by Ames Company of Chicopee, Massachusetts; Miles Greenwood of Cincinnati, Ohio; and William Marshall and Company of St. Louis, Missouri.

As James experimented with different sabot systems and different length projectiles to produce more stabilized flight, his projectile weight-to-bore-diameter ratio exceeded 12 pounds for solid shot, although some 3.8-inch shells weighed 12 pounds.

Nevertheless, the 12-pounder designation stuck for bronze rifled conversions of the Model 1841 6-pounder. And to confuse matters, James came out in late 1861 with his own 3.8-inch bore cannon, a bronze tube manufactured for him by Ames and designated the 14-pounder James rifle.

Both his conversions and his 14-pounder cannon used a 1.25 pound powder charge. The carriages on which they were mounted, the U.S. Model No. 1 of 900 pounds, wouldn't stand the recoil of heavier charges or that produced by heavier projectiles. (Many of these carriages failed in service when used for the even lighter 3-inch U.S. wrought iron rifles [ordnance rifle] and the 2.9-inch or 10-pounder Parrott rifle with one-pound powder charges.)

Light and short James and Schenkl ammunition that fit the 3.67- or 3.8-inch bore cannon and which is 7.5" to 8.5" high (like the original James projectiles) probably were intended for these 12-pounders and 14-pounders and not for the 20-pounder Parrott, which had a 3.67-inch bore but was designed for heavier ammunition, fired two-pound powder charges, and was mounted on the heavier No. 3 carriage weighing 1,175 pounds.

The 3.67- and 3.8-inch rifled bronze guns of the Model 1841 pattern had a bore length of only 57.5". Carriage and tube together weighed only 1,785 pounds. The projectile and powder charge had to be light or the recoil would be excessive and the carriage would fail. (As noted, many did fail and in 1864 the No. 1 carriage design was modified to make it stronger.)

The 20-pounder Parrott, on the other hand, had a bore length of 79", weighed 1,750 pounds, and fired a 20-pound projectile with a two-pound powder charge. But when mounted on the No. 3 carriage, the total weight was 2,850 pounds, more than a half-ton heavier than the 3.67- and 3.8-inch bronze rifles.

This tube and carriage combination could absorb the recoil of the heavier projectile and powder charges, which accounts for the weight and length differences in 3.67-inch projectiles. Although its powder charge and projectile weight was twice as heavy as the 10-pounder Parrott, the 20-pounder had almost exactly the same range at similar tube elevation.

The 3.67-inch projectiles of nearly 20 pounds in weight would increase recoil and strain the carriages to the breaking point if fired in light guns on No. 1 carriages. The light (12 to 14 pounds) projectiles could be fired in 20-pounder guns, but the range tables in the limber chests would be inaccurate as they were calculated for the heavier projectiles.

Western armies and Confederates used a variety of guns throughout the Civil War.

The difference in weight of the two 3.67-inch projectiles was the result of differences in sabots and length. Most intended for 20-pounder rifles by both the North and the South used longer body shells and case-shot to better stabilize the projectile in the longer bore, and featured a short expanding cup sabot.

With the short cup sabot, the solid iron Parrott bolt weighed 19 to 20 pounds. The shells and case-shot were longer, which permitted more room for the explosive charge in shell and more filler balls in case shot. The extra length brought the weight up to the same 19 to 20 pound range to keep all projectiles relatively consistent when fired with the two-pound powder charge. (Range tables pasted in the limber chest contained elevation, range, and time-of-flight information for the three different types of projectiles—solid shot, case shot and shell—so the fuzes could be estimated accurately.)

While Parrott rifles were initially stamped on the muzzle face with both designations (such as "3.67-in. 20 Pdr.") larger rifles came to be designated primarily by bore diameter since weight of shot comparison to smoothbores produced confusion.

The tendency for early collectors to identify any 3.67-inch projectile as intended for "20-pounder rifle" is refuted by the facts. Range tables, powder charges, and projectile weights differed depending on whether the projectile was to be used in a light bronze gun mounted on a No. 1 carriage or in a heavy iron gun (the 20-pounder Parrott, for example) mounted on the No. 3 carriage.

The Federal Army of the Potomac had essentially standardized field artillery batteries by late 1862 by equipping them with Parrott rifles, ordnance rifles, or Napoleons. But the Western armies and the Confederates used a mixture of available guns throughout the war, so it is not unusual that both light and heavy 3.67-inch projectiles are found at late-war battle sites.

—*C. Peter Jørgensen*

Appendix B
Table of Cannon Bore Diameters

This table is presented as an identification tool for matching the cannon with the diameter of the projectiles. It covers field artillery pieces, as well as a few navy and siege artillery pieces. It does not include every known cannon.

BORE DIAMETER		CANNON
1.5	inch	Ellsworth rifle
1.70	inch	Whitworth rifle, 3-pounder
1.75	inch	Confederate infantry gun
2.00	inch	Carkeet gun, 2-pounder
2.125	inch	Woodruff smoothbore, 2-pounder
2.17	inch	Whitworth rifle, 6-pounder
2.25	inch	Confederate mountain rifle
2.32	inch	Pattison rifle, 6-pounder
2.5	inch	Blakely rifle
2.6	inch	Wiard rifle, 6-pounder
2.62	inch	Pattison rifle, 8-pounder
2.75	inch	Whitworth rifle, 12-pounder
2.75	inch	Confederate breech-loading rifle
2.9	inch	Parrott rifle, Model 1861, 10-pounder
2.9	inch	Confederate rifle, 10-pounder
3	inch	Confederate rifle
3	inch	US wrought iron (ordnance) rifle
3	inch	Parrott rifle, Model 1863, 3-inch

BORE DIAMETER		CANNON
3.0	inch	Armstrong rifle, 12-pounder
3.3	inch	Confederate rifle
3.4	inch	Wiard rifle boat howitzer
3.5	inch	Blakely rifle, 12-pounder
3.6	inch	Blakely rifle
3.67	inch	Smoothbore, 6-pounder
3.67	inch	Confederate rifle, 20-pounder
3.67	inch	Delafield rifle, 20-pounder
3.67	inch	Model 1861 rifled 6-pounder
3.67	inch	Parrott rifle, 20-pounder
3.67	inch	Sawyer rifle, 6-pounder
3.67	inch	Wiard rifle, 12-pounder
3.75	inch	Whitworth rifle, 32-pounder
3.8	inch	James rifle, 14-pounder
3.8	inch	Model 1841 James rifled 6-pounder (12-pounder rifle)
4	inch	Blakely rifle, 18-pounder
4.2	inch	Smoothbore, 9-pounder
4.2	inch	Brooke rifle, 30-pounder
4.2	inch	Parrott rifle, 30-pounder
4.4	inch	Dahlgren rifle, 30-pounder
4.5	inch	Siege rifle
4.62	inch	Smoothbore, 12-pounder
4.62	inch	Rifled 12-pounder smoothbore
4.62	inch	Confederate rifled 12-pounder
4.62	inch	James rifle, 24-pounder
4.62	inch	Sawyer rifle
4.62	inch	Mortar, 12-pounder
5	inch	Whitworth rifle, 80-pounder
5.1	inch	Dahlgren rifle, 50-pounder
5.3	inch	Smoothbore, 18-pounder
5.3	inch	Parrott rifle, 60-pounder
5.82	inch	Smoothbore, 24-pounder
5.82	inch	James rifle, 48-pounder
5.82	inch	Sawyer rifle, 48-pounder
5.82	inch	Mortar, 24-pounder

Appendix C
Selected Ordnance Patents

The following list of United States Patents is by no means complete. They are listed here as a starting point for those interested in the technical details of military ordnance. The authors, after exhaustive research, have been unable to locate any detailed Confederate patent material.

NUMBER	DATE	PATENTEE	ITEM PATENTED
12574	3/20/55	W.M.B. Hartley	Making Hollow Shot
12795	5/01/55	E. Hoyt, Jr.	Projectile
12801	5/01/55	W.J. Von Kammerhueber	Projectile
13469	8/21/55	A. McBurth	Shell
13679	10/16/55	A. Hotchkiss	Projectile
13799	11/13/55	S. Sawyer	Shell
14133	1/22/56	W.W. Hubbell	Shell
14315	2/26/56	C.T. James	Projectile
14460	3/18/56	J. Lippincott	Shell-Fuse
14503	3/25/56	W.W. Hubbell	Shell
14742	4/22/56	T. Smith	Projectile
15577	8/19/56	N. Scholfield	Bomb-Lance
15760	9/23/56	A.M. George	Shell
15999	10/28/56	J.B. Read	Projectile
16076	11/11/56	W. Taggart	Projectile
16753	3/03/57	M. Shaw	Projectile
16755	3/03/57	J.M. Sigourney	Projectile
17312	5/19/57	C.C. Brand	Bomb-Lance
17407	5/26/57	R. Sibley	Bomb-Lance
17886	7/28/57	J.L. McConnel	Projectile
17935	8/04/57	T.T.S. Laidley	Projectile For Rifled Cannon
18049	8/25/57	T.T.S. Laidley	Projectile For Rifled Cannon
18401	10/13/57	J.H. Merrill	Projectile
18424	10/13/57	S. Driver	Shell
18707	11/24/57	J.B. Read	Projectile
19505	3/02/58	W.W. Hubbell	Shell
21773	10/12/58	L.B. Olmsted	Shell
23538	5/12/59	W.H. Arnold	Projectile
24371	6/14/59	R. Brown	Bomb-Lance
25951	11/01/59	J.W. Cochran	Projectile
25967	11/01/59	J. Holroyd	Projectile
26016	11/08/59	J.W. Cochran	Projectile
26017	11/08/59	J.W. Cochran	Projectile
26337	12/06/59	J.W. Cochran	Shell
26904	1/24/60	W.W. Hubbell	Shell
27245	2/21/60	B. Swain	Shell
28084	5/01/60	W.W. Hubbell	Projectile

NUMBER	DATE	PATENTEE	ITEM PATENTED
29272	7/24/60	B.B. Hotchkiss	Projectile
29573	8/14/60	Cranston & Bates	Bomb-Lance
29995	9/11/60	W. Rice	Shell
30123	9/25/60	J.W. Cochran	Projectile
31099	1/15/61	S.C. Abbot	Shell
32293	5/14/61	B.B. Hotchkiss	Projectile
32689	7/02/61	J. Gault	Projectile
32702	7/02/61	J. McIntyre	Shell
32844	7/16/61	R. Shaler	Projectile
32949	7/30/61	E. Linder	Shell
32986	8/06/61	J.A. Dahlgren	Projectile
33089	8/20/61	W.F. Ketchum	Shell
33099	8/20/61	R.P. Parrott	Projectile
33100	8/20/61	R.P. Parrott	Projectile
33378	10/01/61	C.F. Brown	Shell
33495	10/15/61	J.P. Schenkl	Shell Fuse
33541	10/22/61	A. Pfund	Projectile
33662	11/05/61	R.P. Parrott	Shell-Fuse
33718	11/12/61	I.P. Tice	Projectile
33746	11/19/61	E. Matteson	Projectile
33754	11/19/61	A.M. Sawyer	Projectile
33835	12/03/61	G.W. Gardner	Shell-Fuse
33863	12/03/61	J.A. Woodbury	Projectile
34031	12/24/61	C.H. Sayre	Projectile
34040	12/24/61	S. Sawyer	Fuse-Hood
34058	1/07/62	B.B. Hotchkiss	Shell
34059	1/07/62	W.W. Hubbell	Shell-Fuse
34091	1/07/62	B.S. Alexander	Projectile For Rifled Ordnance
34206	1/21/62	O. Hopkins	Defending Redoubts
34207	1/21/62	C.T. James	Projectile

NUMBER	DATE	PATENTEE	ITEM PATENTED
34242	1/28/62	Connel & Hall	Shell
34268	1/28/62	J.P. Rollins	Shell
34285	1/28/62	D.M. Mefford	Projectile
34302	2/04/62	G.C. Jones	Shell For Ordnance
34493	2/25/62	W.H. Havens	Projectile
34596	3/04/62	C.W. Small	Projectile
34602	3/04/62	E. Townsend	Shell
34685	3/18/62	J.M. Hathaway	Shell
34693	3/18/62	T.F. Reilly	Projectile
34788	3/25/62	Smith & Stetson	Shell-Fuse
34798	3/25/62	M.S. Wickersham	Shell
34950	4/15/62	C.T. James	Projectile
34965	4/15/62	C.T. James	Projectile
34979	4/15/62	C. Peters	Shell
35103	4/29/62	Mansfield, Morse & Mansfield	Projectile
35153	5/06/62	B.B. Hotchkiss	Shell
35277	5/13/62	C.W. Isbell	Shell
35503	6/10/62	W.E. Browne	Shell
35520	6/10/62	Houghton & Denison	Shell
35521	6/10/62	C.T. James	Projectile
35544	6/10/62	M. Ritner	Sabot For Projectiles
35587	6/17/62	A.B. Cooley	Projectile
35593	6/17/62	J.J. Dresback	Shell-Fuse
35611	6/17/62	B.B. Hotchkiss	Shell-Fuse
35659	6/17/62	Alfred Berney	Improved Composition For Filling
35734	6/24/62	J. Gault	Projectile
35795	7/01/62	R. Weir	Marine Torpedo
35824	7/08/62	J.L. Henry	Projectile For Rifled Ordnance
35825	7/08/62	W.W. Hubbell	Projectile For Rifled Ordnance
35897	7/15/62	J.P. Schenkl	Shell-Fuse
35965	7/22/62	Woodbury & Gray	Sabot For Projectiles
35977	7/22/62	Roys & Lilliendahl	Rocket
36001	7/29/62	T. Goodrem & C. Jackson	Projectile
36037	7/29/62	B.F. Sturtevant	Shell-Fuse
36038	7/29/62	B.F. Sturtevant	Shell-Fuse
36039	7/29/62	B.F. Sturtevant	Shell-Fuse
36116	8/05/62	B.F. Sturtevant	Projectile
36132	8/12/62	J. Absterdam	Coating Projectiles
36172	8/12/62	S. & A.M. Sawyer	Shell-Fuse
36236	8/19/62	J.P. Schenkl	Shell Fuse
36295	8/26/62	W.W. Hanes	Shell
36329	9/02/62	F. Alger	Shell-Fuse
36449	9/16/62	W. Boekel	Projectile
36465	9/16/62	B.B. Hotchkiss	Shell-Fuse
36468	9/16/62	W. Kingsley	Shell-Fuse
36553	9/30/62	F. Alger	Shell-Fuse
36566	9/30/62	W.W. Hubbell	Concussion Fuse For Shells
36664	10/14/62	J. McIntyre	Shell
36686	10/14/62	M.L. Callender	Shell
36773	10/28/62	A.H. Emery	Projectile
36806	10/28/62	W.S. Smoot	Shell-Fuse
36820	10/28/62	S. Groom	Explosive Canister Shot
36858	11/04/62	W.F. Rippon	Shell
36879	11/04/62	J.N. Bird	India Rubber Wad For Projectiles

NUMBER	DATE	PATENTEE	ITEM PATENTED
36934	11/11/62	A. Berney	Shell
37200	12/16/62	S.R. Russell	Shell-Fuse
37260	12/23/62	W.H. Smith	Projectile
37275	1/06/63	J.W. Cochran	Percussion Fuse
37352	1/06/63	H.F. Mann	Projectile
37361	1/06/63	M. Ritner	Sabots And Wads
37520	1/27/63	H. Resley	Shell
37557	1/27/63	C.W. Stafford	Projectile
37565	2/03/63	J.M. Connel	Shell
37566	2/03/63	J.M. Connel	Shell
37661	2/10/63	L.D. Gerardin	Shell
37675	2/17/63	J.W. Cochran	Shell-Fuse
37756	2/24/63	B.B. Hotchkiss	Shell-Fuse
37771	2/24/63	C. Perley	Rocket
37830	3/03/63	J.J. Schillinger	Shell-Fuse
37891	3/17/63	J.B. Atwater	Projectile
37906	3/17/63	A.H. Emery	Projectile
37940	3/17/63	P. Plant	Rocket
37979	3/24/63	B.S. Roberts	Projectile
38322	4/28/63	W.E. Moore	Cartridge
38352	4/28/63	G. Wright	Casting Fuses
38359	4/28/63	C.W. Smith, G.H. Babcock & B.B. & C.A. Hotchkiss	Shell-Fuse
38424	5/05/63	Levi Short	Improved Composition For Filling Shells
38441	5/05/63	L. Wood	Shell
38491	5/12/63	J. Johnson	Shell
38699	5/26/63	A.M. Sawyer	Igniting Time Fuse For Shells
38797	6/02/63	J.L. Henry	Shell-Fuse
38994	6/23/63	I.P. Tice	Shell-Fuse
39112	7/07/63	A. Berney	Projectile
39179	7/07/63	C.W. Stafford	Sabot For Projectiles
39180	7/07/63	C.W. Stafford	Projectile
39216	7/14/63	H.E. Dimick	Projectile
39254	7/14/63	T. Harding	Shell
39267	7/21/63	F. Alger	Shell-Fuse
39282	7/21/63	G.P. Ganster	Projectile
39369	7/28/63	C. Arick	Projectile
39427	8/04/63	C.W. Stafford	Shell
39593	8/18/63	J.N. Smith	Elongated Projectile For Firearms
39612	8/18/63	M.L. Callender	Marine Torpedo
39636	8/25/63	G.H. Felt	Signal Rocket
39682	8/25/63	J.P. Schenkl	Shell-Fuse
39746	9/01/63	J.P. Perry	Explosive Compound For Shells
39897	9/15/63	P.S. Devlan	Improved Composition For Packing Projectiles
39930	9/15/63	L.P. Jenks	Sabot For Hot Shot
39940	9/15/63	W. McCord	Projectile
39942	9/15/63	J. McMurtry	Projectile
39977	9/15/63	C.A. Waterbury	Projectile
40041	9/22/63	J.B. Hyde	Rocket
40054	9/22/63	W. Maginn	Shell
40092	9/29/63	W.H. Dibble	Cartridge
40142	9/29/63	H. Halvorson	Projectile

Selected Ordnance Patents, Continued

NUMBER	DATE	PATENTEE	ITEM PATENTED
40198	10/06/63	C.W. Stafford	Shell
40350	10/20/63	J. McIntyre	Shell-Fuse
40385	10/20/63	O.G. Warren	Projectile
40396	10/27/63	T.H. Burrowes	Shell
40468	11/03/63	S. Gardiner, Jr.	Shell
40538	11/03/63	H. Halvorson	Shell
40828	12/08/63	A.H. Emery	Shell-Fuse
40856	12/08/63	L.E. Reynolds	Projectile
40888	12/08/63	J.N. Smith	Shell
40958	12/15/63	D.E. Somes	Projectile
41112	1/05/64	J.D. Wiloughby	Submarine Explosive Projectile
41127	1/05/64	W.H. Smith	Projectile For Rifled Ordnance
41173	1/05/64	G. Amici	Rocket
41220	1/12/64	H. Helm	Shell
41577	2/09/64	Edward Harrison	Improved Inflammable Composition For Filling Projectiles
41615	2/16/64	Ganster & Schuyler	Shell-Fuse
41668	2/23/64	J. Absterdam	Projectile
41844	3/08/64	T.D. Hart	Explosive Shell
41882	3/08/64	W.S. Williams	Canister Shell
41937	3/15/64	R.P. Parrott	Shell-Fuse
41960	3/15/64	R. Graham	Explosive Shell
42268	4/12/64	W. Boekel	Projectile
42363	4/19/64	G.P. Ganster	Shell-Fuse
42614	5/03/64	W.S. Williams	Projectile
42660	5/10/64	B.B. Hotchkiss	Shell-Fuse
42818	5/17/64	J. Groves	Shell
43017	6/07/64	G.P. Ganster	Attaching Sabot To Spherical Projectiles For Ordnance
43027	6/07/64	B.B. Hotchkiss	Projectiles For Rifled Ordnance
43029	6/07/64	J. Johnson	Explosive Projectile For Ordnance
43801	8/09/64	C.W. Smith	Shell-Fuse
43835	8/16/64	B.C. Converse	Projectile
43922	8/23/64	M. McDevitt	Shell-Fuse
43993	8/30/64	B.B. Hotchkiss	Shell-Fuse
44023	8/30/64	C.W. Stafford	Shell-Fuse
44061	9/06/64	C. Arick	Shell-Fuse
44353	9/20/64	B.H. Tripp	Shell
44546	10/04/64	A. Merryman	Shell-Fuse
44581	10/04/64	J. McIntyre	Shell-Fuse
44588	10/11/64	C. Arick	Shell-Fuse
44669	10/11/64	W.H. Smith	Projectile
44670	10/11/64	W.H. Smith	Projectile
45021	11/15/64	J.F. Cleu	Projectile
45022	11/15/64	J.F. Cleu	Improved Enameled Projectiles
45023	11/15/64	J.F. Cleu	Sabot For Projectile
45024	11/15/64	J.F. Cleu	Shell-Fuse
45035	11/15/64	W.F. Goodwin	Shell-Fuse
45381	12/06/64	G. Wright	Shell-Fuse
45462	12/13/64	J.L. Henry	Projectile
45474	12/20/64	J.F. Cleu	Case-Shot
45567	12/20/64	C.W. Stafford	Sabot For Projectiles
45806	1/10/65	J.S. Adams	Shell-Fuse
45951	1/17/65	F. Schenkl	Packing For Projectiles

Selected Ordnance Patents, Continued

NUMBER	DATE	PATENTEE	ITEM PATENTED
45985	1/24/65	E. Estabrook	Shell
46490	2/21/65	T.G. Orwig	Projectile
47078	3/28/65	C. Arick	Projectile
47213	4/11/65	A.J.S. Molinard	Projectile
47544	5/02/65	B.B. Hotchkiss	Shell
47586	5/02/65	F. Toggenburger	Shell-Fuse
47651	5/09/65	H.W. Libbey	Shell
47725	5/16/65	B.B. Hotchkiss	Projectile
47754	5/16/65	W.H. Smith	Projectile
47776	5/16/65	Wood & Lay	Marine Torpedo
48005	5/30/65	L.E. Williams	Shell
48371	6/27/65	W. Cousins	Projectile
48642	7/11/65	H. Barton	Compound Explosive Shell
48729	7/11/65	C. Sharps	Projectile
49326	8/08/65	L. Wells	Shell
49773	9/05/65	O. Lugo	Cartridge
50357	10/10/65	B.B. Hotchkiss	Projectile
50535	10/17/65	J.H. Hiscock	Shell
50692	10/31/65	E.A. Dana	Sabot For Projectile
50711	10/31/65	W.W. Hubbell	Shell
50783	11/07/65	J. Absterdam	Covering Projectiles
52314	1/30/66	I.P. Noyes	Explosive Shell
55761	6/19/66	L.W. Broadwell	Projectile
55796	6/19/66	S.H. Haycock	Projectile
56712	7/31/66	J.W. Cochran	Projectile
57870	9/11/66	Crane & Fox	Shell
58783	10/16/66	J.M. Currie	Projectile
60567	12/18/66	T.P. Shaffner	Charging Shells
61117	1/08/67	Thomas Taylor	Improvements In Alloys For Sabots Of Projectiles
72494	12/24/67	B.B. Hotchkiss	Combined Time And Percussion Fuse For Explosive Shells
78322	5/26/68	Pertuiset, Mundel & De Fleron	Igniting Explosive Projectiles
82284	9/22/68	C.F. Brown	Projectile

NUMBER	DATE	PATENTEE	ITEM PATENTED
82714	10/06/68	A.O.H. Hardenstein	Explosive Projectile
84635	12/01/68	J. Long	Projectile
87498	3/02/69	R.H. Jones	Projectile
88012	3/23/69	W.A. Cobb	Projectile
88689	4/06/69	J. Absterdam	Projectile
90164	5/18/69	T.D. Gibson	Projectile
90732	6/01/69	J. Curtis	Projectile
91633	6/22/69	J.W. Hill	Projectile
94647	9/07/69	J.D. Richards	Projectile
95137	9/21/69	A.F. Potter	Projectile
96595	11/09/69	J. Jobson	Explosive Projectile
103477	5/24/70	J. Link	Projective
103514	5/24/70	W.H. Shock	Projectile
104108	6/14/70	J.G. Butler	Projectile
107909	10/04/70	J.G. Hope	Projectile
109600	11/29/70	E. Drake	Projectile
112144	2/28/71	B.B. Hotchkiss	Projectile
114807	5/16/71	E.C. Hancock	Improvement In Projectiles
117172	7/18/71	Thomas Hill	Improvement In Projectiles
119313	9/26/71	John G. Butler	Projectile
122014	12/19/71	J.H. Helm	Improved Projectile & Sabot
122620	1/09/72	C. Maduell	Improvement In Projectiles For Small Arms
123828	2/20/72	Samuel E. Jones	Improvement In Projectile For Fire Arms
127796	6/11/72	John Rigney	Improvement In Projectiles
132903	11/12/72	E.A. Dana	Improvement In Sub-Caliber Projectiles
137186	3/25/73	E. Drake	Combined Percussion And Time Fuses For Shells
137442	4/01/73	S.H. Haycock	Projectiles
137491	4/01/73	J.T. Senn	Explosive Shells
138907	5/13/73	F.A. Markley	Compound Explosive Projectiles
142394	9/02/73	J.G. Hope	Projectiles

Selected British Ordnance Patents

Below is a partial listing of the pertinent patents from Great Britain:

604	3/17/55	Bashley Britten	Projectiles
903	4/23/55	Joseph Whitworth	Ordnance, Small Arms, and Projectiles
1740	8/01/55	Bashley Britten	Projectiles For Ordnance
2410	10/29/55	Joseph Whitworth	Artillery and Firearms
1645	6/12/57	Joseph Whitworth	Ordnance, Firearms, and Projectiles
2329	10/19/58	Joseph Whitworth	Guns, Gun Carriages, and Ammunition
1959	8/29/59	Joseph Whitworth	Ordnance, Firearms, and Ammunition
2990	12/30/59	Joseph Whitworth	Projectiles
567	3/01/60	Bashley Britten	Projectiles
585	3/08/61	Bashley Britten	Projectiles For Rifled Ordnance
2104	8/23/61	Joseph Whitworth & William Wilson Hulse	Sights For Small Arms and Ordnance
510	2/25/62	Joseph Whitworth	Projectiles
1663	6/02/62	Joseph Whitworth	Shells

Appendix D
Disarmament and Preservation

We have asked Steve Phillips, a noted professional with over two decades of experience in disarming and preserving Civil War ordnance, to comment on this important area. This is presented as information only. The authors stress that at no time should any attempt be made to disarm any ordnance. Preservation—although not as dangerous—should be done by a professional. Mr. Phillips' remarks follow.

THE PRESERVATION AND HAZARDS OF WAR BETWEEN THE STATES ARTILLERY PROJECTILES

I have been a relic hunter and collector of War Between the States relics for over 20 years now. My interest and specialty has always been artillery projectiles and preservation. This came about because in the beginning I was lucky enough to find a few good projectiles. I was also fortunate enough to meet and become friends with the late Tom Dickey. Without Tom, whose extensive work and interest in projectiles educated us all, we would know very little about most of the more obscure types. Most of those who have been involved with artillery projectile collecting are directly or indirectly tied to the hobby because of Tom Dickey.

Please: Do not attempt to defuse a live shell yourself. Get an expert.

A consequence of my collecting efforts has been that I have learned many ways to defuse shells and to preserve iron, wood, and leather relics with a good degree of success. In that time, I have experimented with a wide variety of preservation techniques and have observed many other collectors and museums in their efforts. If have found that there is no absolute one way to preserve relics: all methods are, to some degree, experimental. Unfortunately, many techniques fail utterly to stop the seemingly inevitable deterioration of iron projectiles, and yet they are still in use today by both individuals and museums. The techniques and ideas that I will be discussing are those that I have found to be successful in drastically slowing or totally stopping the deterioration of iron relics.

First and foremost, do not attempt to defuse a shell yourself. Further, unless you have incontrovertible proof that a shell has been properly unloaded, always assume that it is live. *These shells can and do explode.* They will explode underwater, and they will explode unexpectedly when being drilled. The only attitude to take when unloading shells is to assume that they are going to explode while drilling, whether underwater or not. The black powder inside a War Between the States projectile does not go bad inside the shell; in fact, in some cases it is more potent today than when it was manufactured. Within the last ten years there have been at least six instances in which shells have exploded while being unloaded.

The only way to unload a live shell is by remote rig, and this work should be left to professionals. Even shells that appear to have been unloaded can sometimes pose a threat if the unloading was done improperly. Air pockets are often present inside shells, and these can cause damage similar to a powder explosion. There are collectors out there who know how to properly unload a shell, and you should contact them for this service. Don't risk your life and limbs over an old rusty piece of scrap iron. I feel certain that we have yet to see the last casualty of the war: it will result from amatuer efforts to unload a live shell.

The application of polyurethane or paint

has always been a common method of projectile preservation. This is only slightly better than no preservation at all, and I do not recommend applying either substance. If polyurethane works well on a shell, then that shell probably does not need any preservation at all. Polyurethane will temporarily provide a hard outer casing on a shell, but it also traps moisture inside. While the shell may look great in the beginning, polyurethane does nothing to actually preserve the shell and in most cases it will continue to deteriorate from the inside out. After some years have passed, your carefully polyurethaned shell will probably fall apart in large chunks. In short, it is not recommended for long-range preservation.

Along this same line is the use of paint. Many otherwise highly collectible shells have been painted black in order to mimic their original appearance. This does nothing to preserve the shell and is frowned upon by serious collectors, since it gives the impression that the shell's owner may be trying to hide some repair or otherwise misrepresenting the facts.

These techniques have proved effective in slowing or stopping deterioration.

There are basically two steps to take in order to preserve a War Between the States projectile: stabilization and preservation. Stabilization is accomplished through the use of electrolysis, of which there are two types: fast mechanical cleaning and slow sulfide-removal cleaning. Mechanical cleaning is really just an extremely fast version of the slower type of electrolysis, wherein the sulfides are extracted from the shell and the outer coating of rust and oxidation is removed. Mechanical cleaning uses a higher current, whereas slow electrolysis uses low current. Mechanical electrolysis is usually more of an overnight process; slow electrolysis can often take weeks or more. Generally, slow electrolysis is used on large objects or more heavily oxidized and damaged relics. The state of Florida currently has some Spanish cannons that will stay in an electrolysis bath for years!

Some projectiles will be fine with little or no preservation work. Others are extremely difficult to stabilize due to a number of different factors. Waterlogged wood sabots and fuzes can present the most difficult preservation problems, and they will shrink and fall apart if not properly dealt with. Some other factors include: the location found, lost body weight due to a natural form of electrolysis, metals leaching out, exposure to fertilizers and chemicals, close proximity to other shells, and exposure to water. Saltwater shells in particular are difficult to stabilize. Any brackish or saltwater shells from the coast are going to present major problems in terms of both stabilization and preservation. The deep metal sulfides in these shells should not be cleaned with a mechanical electrolysis, which would leave only an unsightly rough core. They can generally be stabilized with slow electrolysis and other methods of sulfide removal and then impregnated with a micro-crystalline wax treatment. This wax treatment—when done properly—is a complicated and extensive process taking many days or even weeks to complete. This is not a wax "dipping," which would be unappealing and useless. What a proper wax treatment accomplishes is total permeation of the shell, both on the surface and below, with a solid barrier to any further decomposition or environmental effects.

Unfortunately, wax treatments pose many hazards. Some types of wax, such as paraffin, can be extremely flammable. The air pockets already mentioned can become explosive if subjected to excessive heat. Lead sabots and fuzes can melt and be ruined, and there is always the chance that you could spill a huge vat of semi-boiling wax on yourself or others. I have personally had expanding air pockets blow out two shells, sending gallons of deadly melted wax out in all directions. Again, this is not an operation to be taken lightly nor without the proper advice and assistance.

Another factor any serious collector

should consider is after-the-fact environmental controls. Even a well-preserved shell will continue to deteriorate if exposed to environmental extremes. Following preservation, projectiles should be kept in a fairly constant environment, such as inside your home where extremes in temperature and humidity are less likely occur. Many collectors, particularly those in normally humid parts of the country like myself, also use a dehumidifier in order to extract excess moisture from the room where the projectiles are displayed. Controlling these two factors—heat and humidity—will slow deterioration and, coupled with a good preservation process, will keep your projectiles in excellent condition indefinitely.

Many people ask me, "Why cannonballs?" As I mentioned earlier, I was fortunate to find several of them when I first became interested in War Between the States artifacts. In addition, the wide variety of styles and sizes has helped sustain my interest over the years. Further, projectiles are hard to fake and heavy to steal.

Another reason my interest in projectiles has endured to this day is because they were the deadliest weapons used in the war. Please keep in mind that they can still be deadly today.

I hope that all collectors will take care in stabilizing and preserving their projectile collections for future generations to study.

—*Steve Phillips*

Appendix E
Recommended Readings

The following list presents the interested reader with a number of useful resources. Most of the books noted are in print and readily available to the collector.

BOOKS

Artillery and Ammunition of the Civil War by Warren Ripley. New York: Van Nostrand Reinhold Co., 1970.

Cannons: An Introduction to Civil War Artillery by Dean Thomas. Gettysburg, Pa.: Thomas Publications.

Civil War Projectiles II: Small Arms & Field Artillery, With Supplement by W. Reid McKee and M.E. Mason, Jr. Originally printed 1980. Reprint Mechanicsville, Va.: Rapidan Press.

Civil War Relics of the Western Campaign, 1861-1865 by Charles S. Harris. Mechanicsville, Va.: Rapidan Press, 1987.

Field Artillery Projectiles of the American Civil War, 1980 edition by Thomas S. Dickey and Peter C. George. Atlanta: Arsenal Press, 1980.

A Field Guide for Civil War Explosive Ordnance by John D. Bartleson, Jr. Washington: U.S. Government Printing Office, 1972.

The Illustrated History of American Civil War Relics by Stephen W. Sylvia and Michael J. O'Donnell. Orange, Va.: Moss Publications, 1978.

North South Trader's Civil War Collector's Price Guide edited by Nancy Dearing Rossbacher. Orange, Va.: Publisher's Press, Inc.; 6th edition, 1993.

Shiloh Shells and Artillery Units by George F. Witham. Memphis, Tenn.: Riverside Press, 1980.

PERI0DICALS

North South Trader's CIVIL WAR. An illustrated bi-monthly published in Orange, Va., by Publisher's Press, Inc.; specifically designed for Civil War collectors and relic hunters.

The Artilleryman. Published quarterly in Tunbridge, Vt., by Cutter & Locke, Inc.; articles of interest to the muzzleloading enthusiast.

Civil War News. A newspaper-format periodical published ten times yearly in Tunbridge, Vt., by Cutter & Locke, Inc.; features current events in the Civil War field.

Civil War Times Illustrated. Published bi-monthly in Mt. Morris, Il.; the oldest of the general-topic Civil War periodicals.

America's Civil War. Published bi-monthly in Mt. Morris, Il.; similar in format to *CWTI,* above.

Blue & Gray. Published bi-monthly in Columbus, Oh.; commonly devotes an entire issue to a specific action or battlefield.

Military Images. Published bi-monthly in Henryville, Pa.; presents the images of war, with a concentration on Civil War.

The Courier. Newspaper format bi-monthly published in Williamsville, N.Y.; contains ads, queries, book reviews, and upcoming events.

Note: Many relic hunting associations also produce informative bulletins and newsletters.

Bibliography

Abbot, Henry L. *Siege Artillery in the Campaigns Against Richmond.* New York: D. Van Nostrand, 1868.

Bartleson, John D. *Civil War Explosive Ordnance 1861-1865.* Washington: U.S. Government Printing Office, 1972.

Description of the Whitworth System of Rifling, Also of Rifled Ordnance and Ammunition. Manchester: George Falkner, 1867.

Dew, Charles B. *Ironmaker to the Confederacy.* Wilmington, North Carolina: Broadfoot Publishing Company, 1987.

Dickey, Thomas S., and Peter C. George. *Field Artillery Projectiles of the American Civil War,* 1980 edition. Atlanta: Arsenal Press, 1980.

Gibbon, John. *The Artillerist's Manual, Etc.* New York: D. Van Nostrand, 1863.

Gillmore, Q.A. *Engineer and Artillery Operations Against the Defenses of Charleston Harbor in 1863.* New York: D. Van Nostrand, 1865.

Hazlett, James C., Edwin Olmstead, and M. Hume Parks. *Field Artillery Weapons of the Civil War.* Newark: University of Delaware Press, 1983.

Howard (Senator, initials unstated). *Reviews of Dr. J.B. Read's Improvements and Discoveries . . . Of Projectiles for Rifles Ordnance.* Tuscaloosa: W.H. Sugg, 1884.

Jones, J. William. *Southern Historical Society Papers,* Volume XI. Millwood, New York: Kraus Reprint Company, 1977.

Kerksis, Sydney C., and Thomas S. Dickey. *Field Artillery Projectiles of the Civil War, 1861-1865.* Atlanta: The Phoenix Press, 1968.

Kerksis, Sydney C., and Thomas S. Dickey. *Heavy Artillery Projectiles of the Civil War, 1861-1865.* Atlanta: The Phoenix Press, 1972.

McKee, W. Reid, and M.E. Mason, Jr. *Civil War Projectiles II, Small Arms and Field Artillery, with Supplement.* Mechanicsville: Rapidan Press, 1980.

Paulding, J.N. *The Cannon and Projectiles Invented by Robert Parker Parrott.* New York, 1979.

Peterson, Harold L. *Round Shot and Rammers.* Harrisburg: Stackpole Books, 1969.

Ripley, Warren. *Artillery and Ammunition of the Civil War.* New York: D. Van Nostrand Reinhold Co., 1970.

Stevens, Phillip H. *Artillery Through the Ages.* New York: Franklin Watts, Inc., 1965.

The Confederate Field Manual. Richmond: Ritchie and Dunnavant, 1862.

The War of the Rebellion (The Official Records). 70 volumes bound in 128 parts and a two-part Atlas. Washington: U.S. Government Printing Office, 1880-1901.

Index

from Concussion Fuse to Hotchkiss & Sons

About the Authors

Jack W. Melton, Jr.

Jack W. Melton, Jr., a life-long resident of Georgia, was born in Columbus on May 10, 1960. Both geography and heredity destined him to be a student of the Civil War: he grew up adjacent to the Kennesaw Mountain National Military Park near the Kolb Farm, several of his ancestors fought for the South during the war (his great-grandfather was an artillery driver for Milton's Light Artillery), and both of his parents are collectors of Civil War artifacts.

His interest in the history-rich area surrounding him spurred him to respond to the National Park Service's need for a historical trail through their Kennesaw park. This work became his Eagle Scout project, and he became a Bicentennial Eagle Scout on Independence Day 1976. Since that time, thousands of people have walked the trail he cut at Kennesaw Mountain.

He began relic hunting at age 14, and found his first 12-pounder solid shot that same year. Thus began his interest in the field artillery projectiles of the War Between the States. In the ensuing years, his interest in excavated artillery projectiles has taken him from Port Hudson to Gettysburg, and he has recovered nearly 250 projectiles.

Deeply affected by the death of renowned artillery collector and author Thomas Dickey, Sr., in 1987, Jack was moved to continue the work of the man he so admired, researching, photographing, and studying the wide variety of projectiles and their patents.

He entered the realm of professional Civil War dealers when he purchased a collection in 1988. He discovered that he had an affinity for buying and selling artifacts, and he opened his first shop in 1989. He also began to pursue professional photography, and his work has appeared within the pages and on the front cover of several national publications. He is a featured front cover photographer for *North South Trader's CIVIL WAR* magazine.

Today he owns and operates Kennesaw Mountain Military Antiques in Kennesaw, Georgia. He is also currently in the professional photography program at the Southeastern Center for the Arts in Atlanta.

Lawrence E. Pawl

Lawrence E. Pawl was born December 6, 1948, in Grand Rapids, Michigan. He completed his undergraduate work at Michigan's Grand Valley State University and was graduated with a degree in Health Sciences. He went on to obtain his medical degree at Wayne State Medical School in Detroit.

His interest in the Civil War dates back to age 14, when he attended a Civil War show in Detroit. He purchased his first artillery shell at age 25—a 3.67-inch Hotchkiss fired at Shiloh—and his fascination with the projectiles of the Civil War was born.

In the late 1970s, he encountered two books that were to have a profound effect on him. In an it's-a-small-world coincidence, Grand Rapids medical colleague (and now partner) Jack O'Donnell introduced him to *The Illustrated History of American Civil War Relics,* written by O'Donnell's brother Mike and Stephen W. Sylvia. He then encountered a copy of Dickey and George's *Field Artillery Projectiles of the American Civil War* and began to correspond with and purchase projectiles from Tom Dickey. He recalls that Tom often shipped the weighty shells cross-country to him via Greyhound bus.

A mutual interest in artillery projectiles brought together co-authors Pawl and Melton in the mid-1980s. This volume is the result of their shared love of research on the surviving projectiles and their patents.

Dr. Pawl has been in private practice in Grand Rapids since 1981, specializing in medical oncology. In his leisure hours, of which there aren't nearly enough, he also collects excavated Civil War weaponry, including swords, longarms, and handguns. His collecting interests also go as far afield as astronaut memorabilia, particularly items relating to Apollo 11 and Buzz Aldrin.

He has two teenage children: son Andy (who, much to his father's delight, is showing some interest in the War Between the States) and daughter Sarah.

Center: co-authors Jack W. Melton, Jr. (left) and Lawrence E. Pawl at the 1993 Middle Tennessee Civil War Show & Sale in Nashville, Tennessee. Photo by Stephen W. Sylvia.